General Certificate of Secondary Education

GCSE
Science: Double Award (Coordinated)
Paper 1 – *Higher Tier*

Centre name					
Centre number					
Candidate number					

Physics

Surname
Other names
Candidate signature

Time allowed: 1 hour 30 minutes.

Instructions to candidates
- Write your name and other details in the spaces provided above.
- Answer **all** questions in the spaces provided.
- Do all rough work on the paper.
- Write your answers in black or blue ink or ball-point pen.

Information for candidates
- The marks available are given in brackets at the end of each question or part-question.
- Marks will not be deducted for incorrect answers.
- In calculations show clearly how you work out your answers.
- State the units in all your answers.
- There are 13 questions in this paper. There are no blank pages.

Advice to candidates
- Work steadily through the paper.
- Don't spend too long on one question.
- If you have time at the end, go back and check your answers.

For examiner's use

Q	Attempt Nº 1	2	3	Q	Attempt Nº 1	2	3
1				8			
2				9			
3				10			
4				11			
5				12			
6				13			
7							
				Total 100			

© CGP 2002

1 The Highway Code gives tables of the shortest stopping distances for cars travelling at various speeds. An extract from the Highway Code is given below.

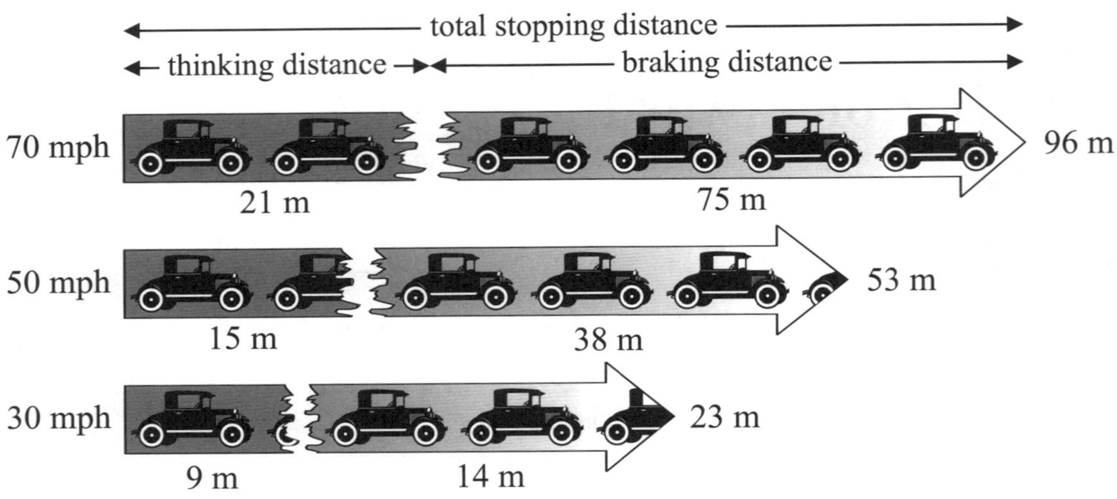

(a) (i) How does the thinking distance change with speed?

....

(1 mark)

(ii) Why isn't the thinking distance the same for all three speeds?

....

(1 mark)

(b) Write down **two** factors other than speed that would affect the thinking distance.

1. ....

2. ....

(2 marks)

(c) Explain what would happen to the braking distance if the car was full of people.

....

....

....

(2 marks)

Leave blank

© CGP 2002

(d) A racing car was travelling at 45 m/s when the driver applied the brakes. The graph below is a velocity-time graph showing the velocity of the car during braking.

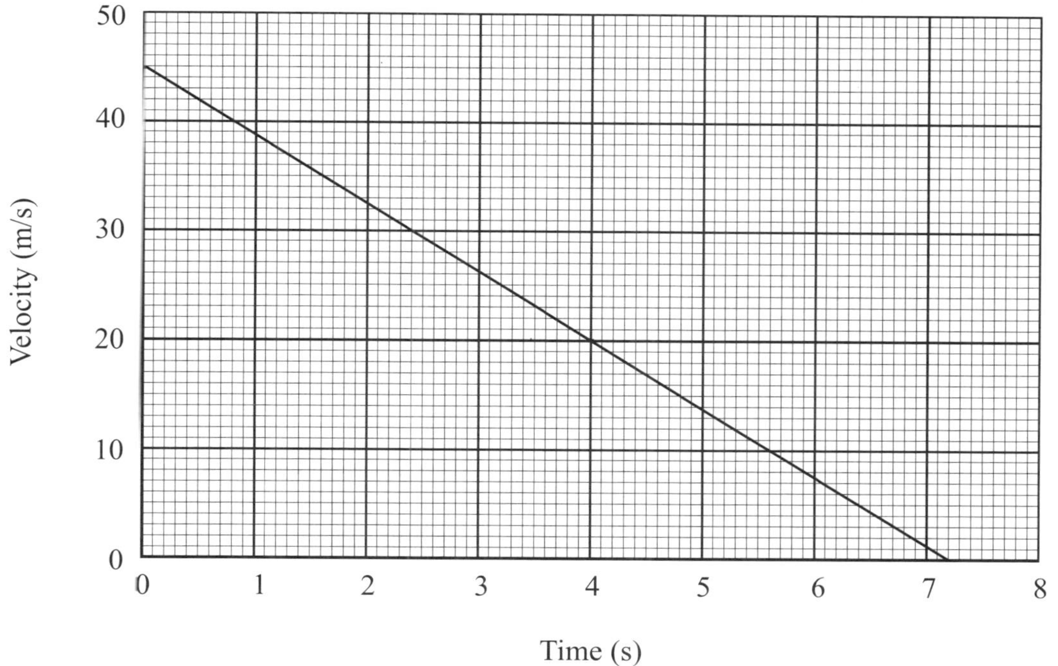

(i) What is the car's speed 5 seconds after the brakes are applied?

....

(1 mark)

(ii) Calculate the rate at which the velocity decreases (deceleration).

....

....

Rate = ... *m/s²*

(2 marks)

(iii) Calculate the braking force if the mass of the car is 600 kg.

....

....

braking force = ... *N*

(2 marks)

(iv) Calculate the braking distance.

....

....

braking distance = ... *m*

(2 marks)

Leave blank

© CGP 2002

Leave blank

2 This question is about our universe.

(a) Look at this list.

solar system **planet** **comet** **galaxy** **star**

(i) Which is the largest? Choose from the list.

....

(1 mark)

(ii) Which is the smallest? Choose from the list.

....

(1 mark)

(b) The Sun in our solar system is a star just like those we see at night.

(i) Explain why we can't see other stars during the day.

....

....

(1 mark)

(c) Stars have very distinct stages in their lives.
This is a list of stages in the life of a small star like the Sun.

1) Clouds of dust and gas contract and start to fuse.
2) This forms what's called a 'protostar' which develops into a main sequence star.

After this stable period of a star's life, some major changes take place.
Describe and explain the changes that take place as a star changes from a main sequence star to a white dwarf.

....

....

....

....

(3 marks)

© CGP 2002

3 Here is a diagram of the apparatus needed to copper-plate a lead model.

Leave blank

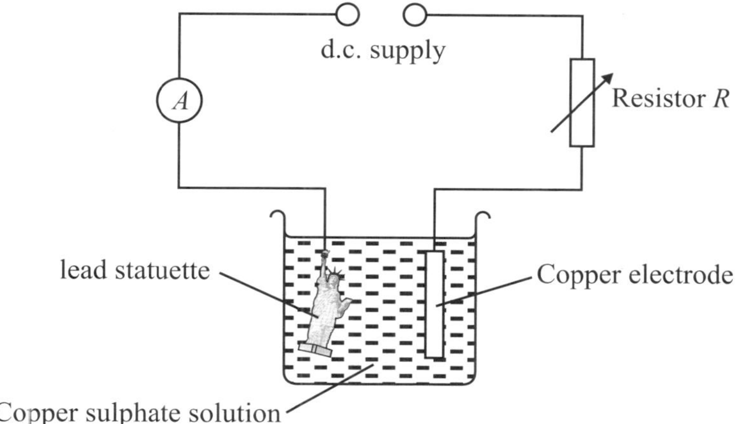

Electricity travels through the liquid as a flow of ions.
Copper sulphate solution contains copper ions which are positively charged.
Copper sulphate solution conducts electricity.

(a) (i) Which terminal of the d.c. supply should the model be connected to?

....

(1 mark)

(ii) Give a reason for your answer.

....

....

(2 marks)

(b) The quality of the copper-plating is improved if a lower current is passed through the solution for a longer time.
Give **two** ways in which the circuit could be altered to reduce the current.

1. ....

....

2. ....

....

(2 marks)

© CGP 2002

Leave blank

4 The diagram shows part of a bicycle wheel and a dynamo attached to it.
When the bicycle wheel turns, a current flows through the wires and causes the bulb in the headlamp to light.

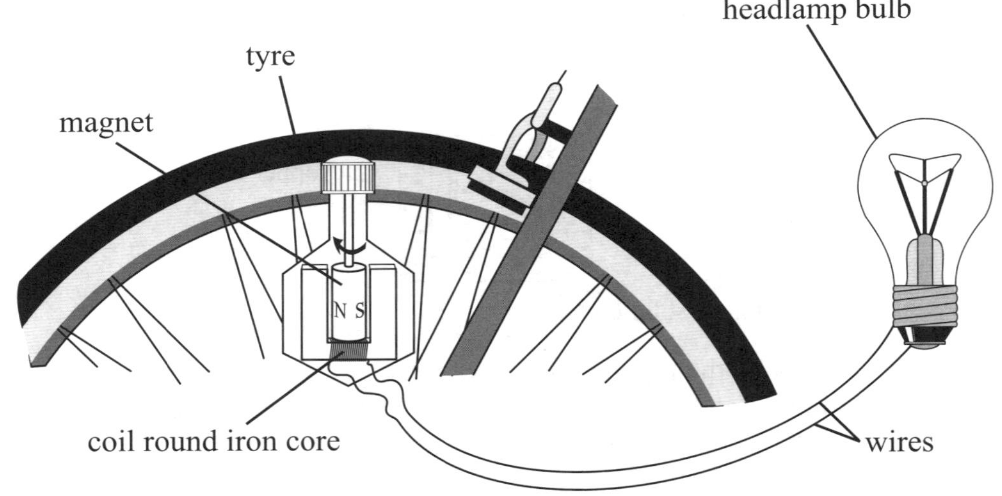

(a) Explain, as fully as you can, how the current is generated.

....

....

....

....

....

(3 marks)

(b) Is the output from this dynamo alternating current or direct current? Explain why.

....

....

(2 marks)

(c) The bulb glows brighter when the voltage across it is increased.
Give **three** ways in which the bulb could be made to glow brighter.

1. ....

2. ....

3. ....

(3 marks)

© CGP 2002

5 Slate is transported from a hillside quarry to the bottom of the hill by a small railway cart. The cart is lowered down the slope and then pulled back up by a long cable. The cable passes around pulleys at the top and bottom of the slope. The pulleys are driven by an electric motor.

Leave blank

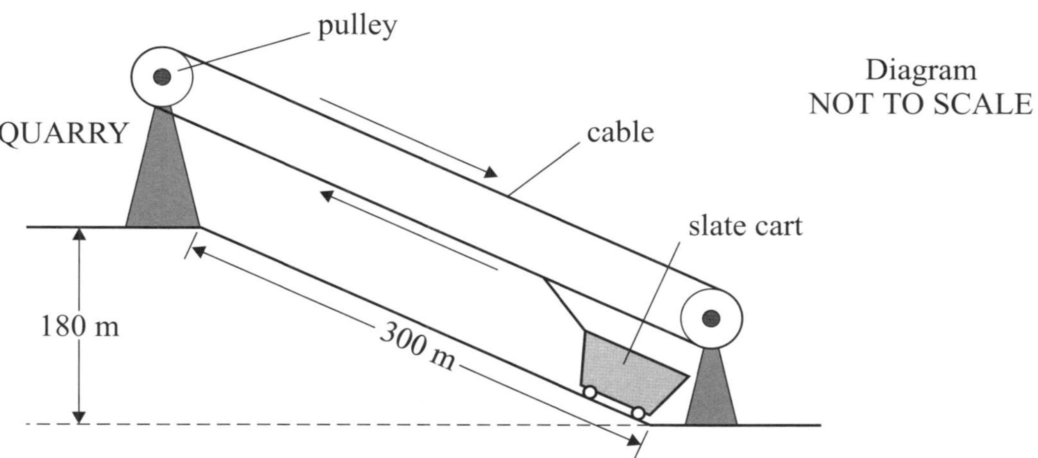

(a) The empty cart weighs 6500 N.
The track is 300 m long and rises a vertical height of 180 m.

(i) Calculate the useful work done by the motor just to lift the empty cart back to the top of the track. Give your answer in kJ (kilojoules).

....

....kJ

(3 marks)

(ii) The cart takes 4 minutes to travel up the slope.

Use $\text{Power} = \dfrac{\text{Work done}}{\text{Time taken}}$ to calculate the power needed just to lift the empty cart back to the top of the track. Don't forget to include the units.

....

power: ....

(3 marks)

(b) The power output of the motor is greater than the power needed just to lift the empty cart back to the top of the track.
Give **two** reasons for this.

1. ....

2. ....

(2 marks)

© CGP 2002

6 Carbon-14 is found in all living things. When living things die, the carbon-14 begins to decay gradually. Its half-life is 5 600 years.

(a) Explain what is meant by the term 'half-life'.

....

....

(1 mark)

(b) A human skeleton was found in an ancient burial ground.
Explain how carbon dating could be used to find the age of the skeleton.

....

....

(2 marks)

(c) Carbon-14 makes up about one ten-millionth of the carbon in the air and in living things.

(i) When the skeleton was dug up, it was found that only one out of every forty million carbon atoms in the bone was carbon-14.
For how many half-lives of carbon-14 has the skeleton been in the ground?

....

....

....

....

(3 marks)

(ii) Approximately how old is the skeleton?

....

(1 mark)

(d) The skeleton was wearing a necklace made of small stones.
Why is it not possible to carbon-date this necklace?

....

(1 mark)

(e) Another skeleton was found to have one part carbon-14 in every 20 million parts carbon. Did this skeleton die before or after the original skeleton?

....

(1 mark)

Leave blank

© CGP 2002

Leave blank

7 (a) Two rays of light are shone through two blocks of glass.
The diagram shows the paths these rays take through the blocks of glass.

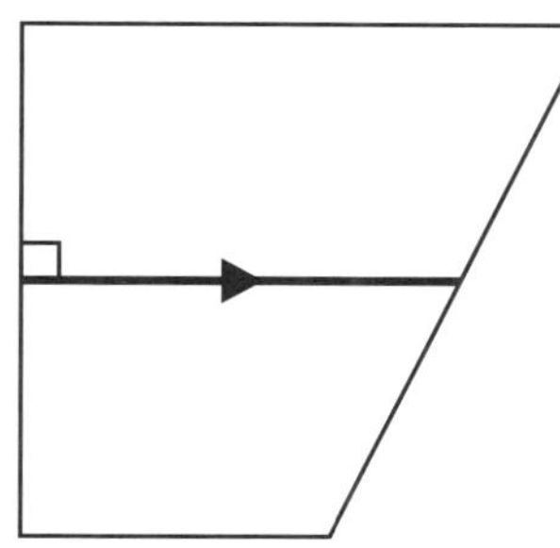

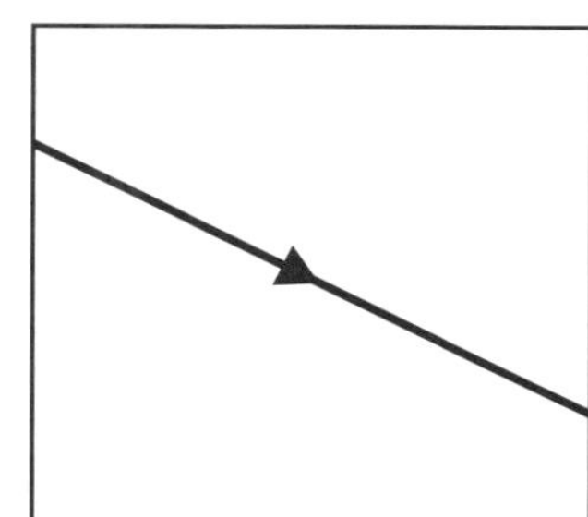

(i) On the diagram above, draw the path each ray would follow through the air before and after passing through the blocks of glass.

(2 marks)

(ii) Find two pairs of angles along the second path that are equal, and label them X and X´, and Y and Y´.

(2 marks)

(b) A ray of light is shone through a semicircular piece of glass from two different angles, as shown in the diagrams below.

Diagram 1

Diagram 2

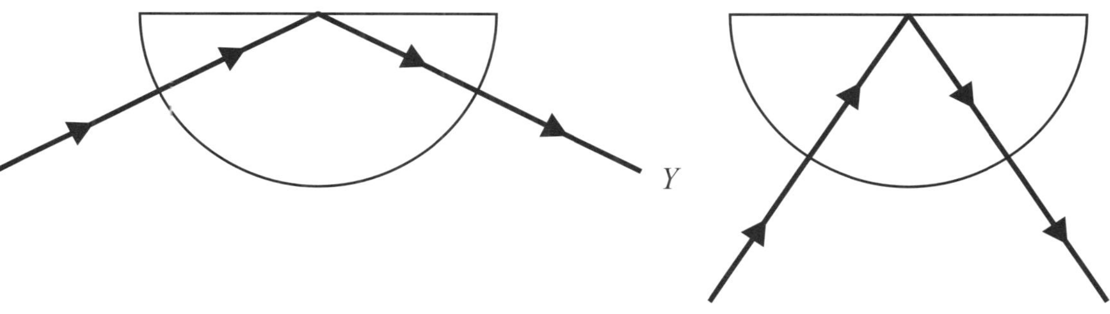

Explain why the reflected beam viewed from *Y* is brighter than the reflected beam viewed from *Z*. Add to the diagram if it helps your answer.

....

....

....

....

(3 marks)

© CGP 2002

Leave blank

8 The diagram shows two satellites orbiting the Earth.
One is in a geostationary orbit, and the other is following a polar orbit.

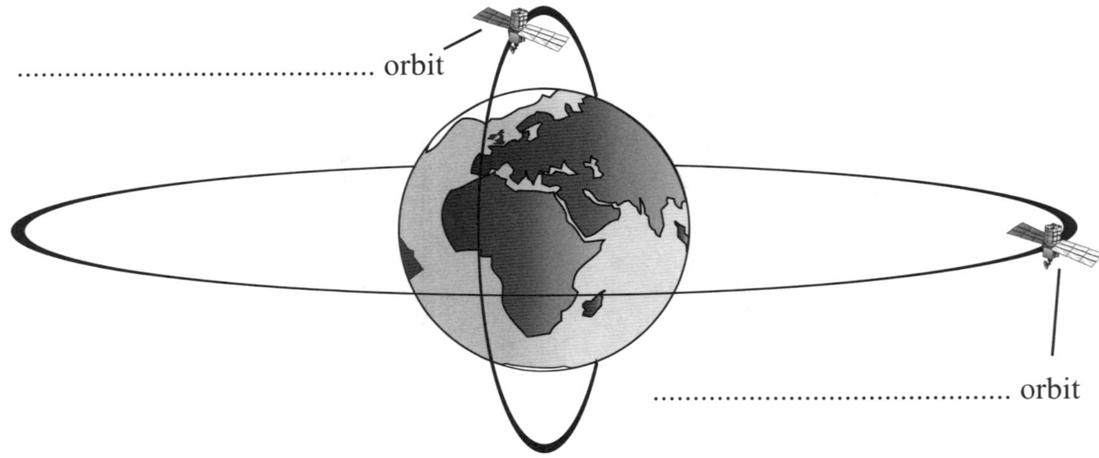

(a) (i) Label the two orbits in the diagram.

(1 mark)

(ii) The orbital time for the geostationary satellite is 24 hours.
How is that of the polar satellite different?
Explain your answer.

....

....

(2 marks)

(b) Give **two** reasons why a polar orbit is better for a spy satellite.

1. ....

....

2. ....

....

(2 marks)

(c) Suggest **one** possible use for a satellite in a geostationary orbit.

....

....

(1 mark)

© CGP 2002

Leave blank

9 (a) When Andrew gets hot, his skin produces sweat, which evaporates.
Explain how this helps to keep Andrew cool. Use ideas about particles in your answer.

....

....

....

....

....

(3 marks)

(b) Andrew and Adam go for a walk on a sunny day. They take bottles of water with them.
Andrew's water bottle is a shiny silver colour. Adam's is black.

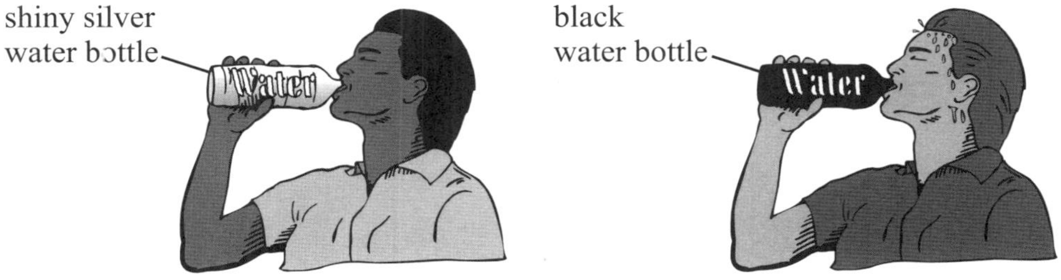

shiny silver water bottle

Water

black water bottle

Water

Andrew **Adam**

Explain why Andrew's water stays cooler than Adam's.

....

....

(2 marks)

(c) Adam's house is insulated. Name **two** methods of insulating a house and for each method say how this could reduce heating bills.

1 Type of insulation: ....

How it works: ....

....

2 Type of insulation: ....

How it works: ....

....

(4 marks)

© CGP 2002

Leave blank

10 (a) (i) Sound waves are longitudinal. Explain what **longitudinal** means.

....
(1 mark)

(ii) What is ultrasound?

....
(1 mark)

(iii) Sound waves travel faster through water than they do through air.
What would happen to the frequency and the wavelength of a sound wave as it passed from air to water?

....

....
(2 marks)

(b) A small speaker which produces ultrasonic waves is placed a short distance from a large sheet of metal that has a small gap in it.

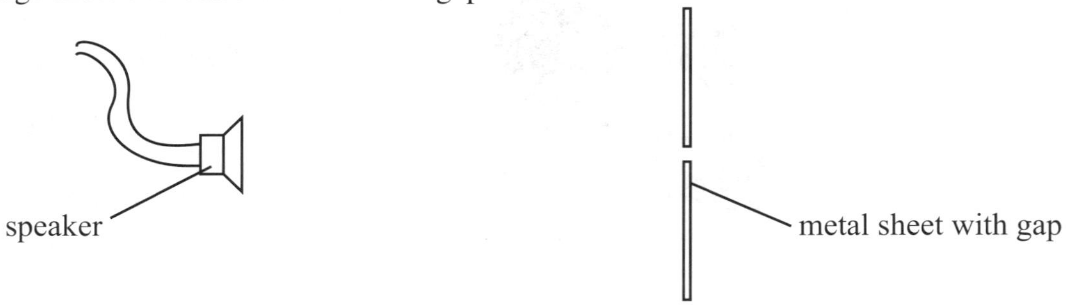

speaker

metal sheet with gap

An oscilloscope is set up with a small microphone to detect ultrasonic waves.

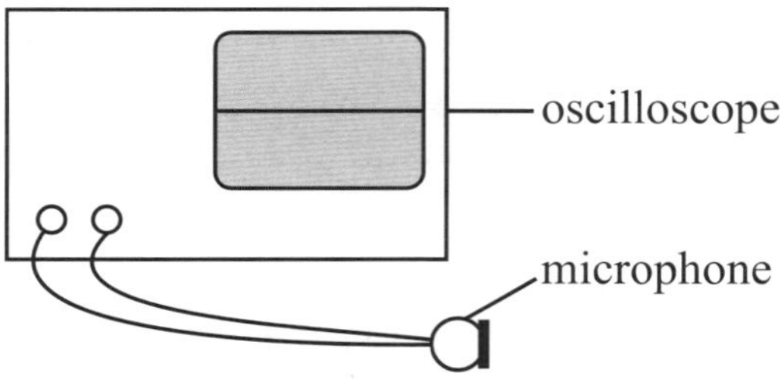

oscilloscope

microphone

ultrasound not detected

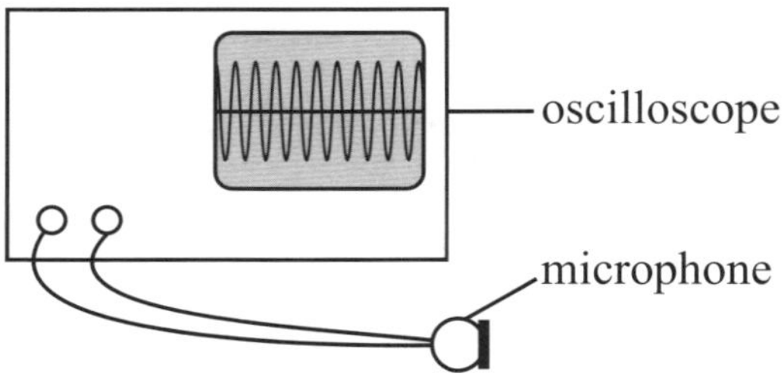

oscilloscope

microphone

ultrasound detected

Explain how you could use this apparatus to demonstrate that the ultrasonic waves from the speaker are being diffracted.

....

....

....

....
(3 marks)

© CGP 2002

Leave blank

(c) Scientists use pulses of ultrasound to investigate deep, murky water.

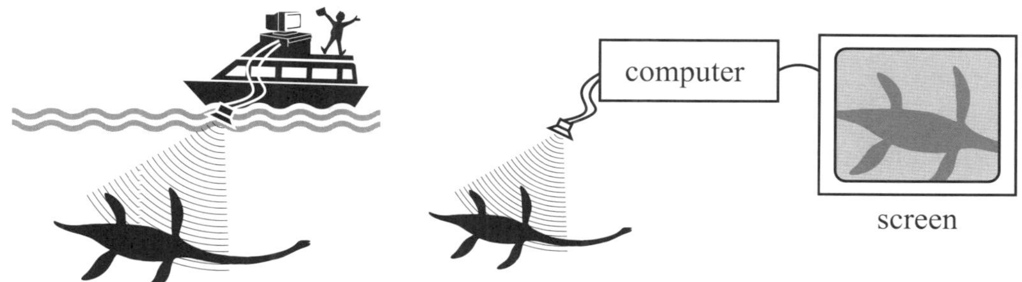

The waves enter the water via a transmitter on the bottom of a boat. Any changes in the density of the water, such as solid objects, algae or fish, will reflect the ultrasonic waves back to the transmitter.
The transmitter also acts as a receiver, and is attached to a computer which calculates how far each pulse travelled in the water.
The computer generates a picture of what is found under the surface of the water.
Explain the advantages of ultrasound over X-rays to examine the contents of the lake.

....

....

....

....

(3 marks)

Turn over for next question.

© CGP 2002

Leave blank

11 The diagram below shows a fire alarm.
When the glass is broken, the spring closes the switch and the alarm goes off .

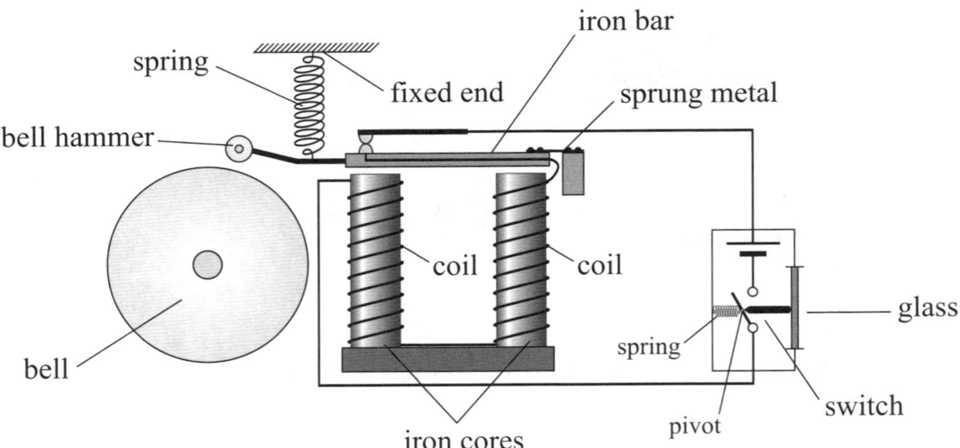

Explain how the bell is rung when the glass is broken.

....

....

....

....

(4 marks)

© CGP 2002

Leave blank

12 (a) Atoms are made up of three types of particle called protons, neutrons and electrons. Complete this table to show the relative mass and charge of a neutron and an electron.

Particle	Relative mass	Relative charge
proton	1	+1
neutron		
electron		

(2 marks)

(b) The diagram below shows five different nuclei.

A B C D E

● — proton
○ — neutron

Which two are isotopes of the same element?

.... and ....

(1 mark)

Explain your answer.

....

....

(1 mark)

(c) Some atoms have an unstable nucleus which can decay and emit nuclear radiation.

(i) Why is it necessary to take precautions when handling sources of nuclear radiation?

....

....

(1 mark)

(ii) Describe **two** safety measures that might be taken by a person working with sources of nuclear radiation.

1. ....

2. ....

(2 marks)

© CGP 2002

Leave blank

13 A mintball sweet is placed on the surface of some fairly runny honey in a jar.
When the mintball is released it falls through the honey.
The diagram on the left shows the forces P and Q acting on the mintball as it falls.

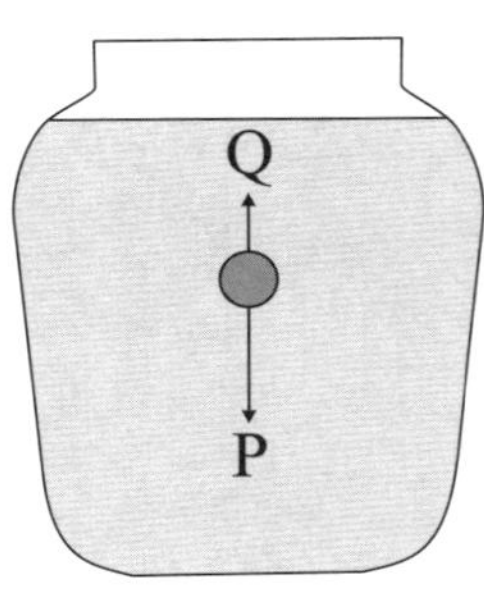

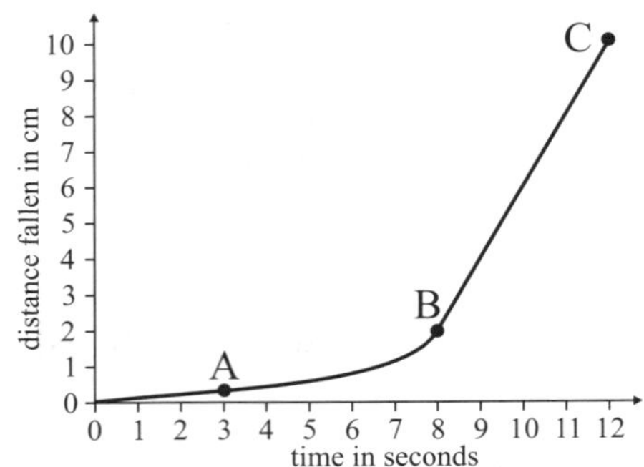

The graph on the right shows how the distance the mintball has fallen varies with time.

(a) (i) What happens to the speed of the mintball between points A and B?

....

(1 mark)

(ii) Explain why this happens by referring to forces P and Q.

....

....

....

(1 mark)

(b) (i) What happens to the speed of the mintball after point B?

....

(1 mark)

(ii) Explain why this happens by referring to forces P and Q.

....

....

....

(3 marks)

(c) Use the graph to calculate the speed of the mintball between points B and C.

....

(2 marks)

PHP4U

© CGP 2002

General Certificate of Secondary Education

GCSE
Science: Double Award
(Coordinated)
Paper 2 – *Higher Tier*

Centre name					
Centre number					
Candidate number					

Physics

Surname
Other names
Candidate signature

Time allowed: 1 hour 30 minutes.

Instructions to candidates
- Write your name and other details in the spaces provided above.
- Answer **all** questions in the spaces provided.
- Do all rough work on the paper.
- Write your answers in black or blue ink or ball-point pen.

Information for candidates
- The marks available are given in brackets at the end of each question or part-question.
- Marks will not be deducted for incorrect answers.
- In calculations show clearly how you work out your answers.
- State the units in all your answers.
- There are 13 questions in this paper. There are no blank pages.

Advice to candidates
- Work steadily through the paper.
- Don't spend too long on one question.
- If you have time at the end, go back and check your answers.

For examiner's use

Q	Attempt Nº 1	2	3	Q	Attempt Nº 1	2	3
1				8			
2				9			
3				10			
4				11			
5				12			
6				13			
7							
				Total 100			

© CGP 2002

Leave blank

1 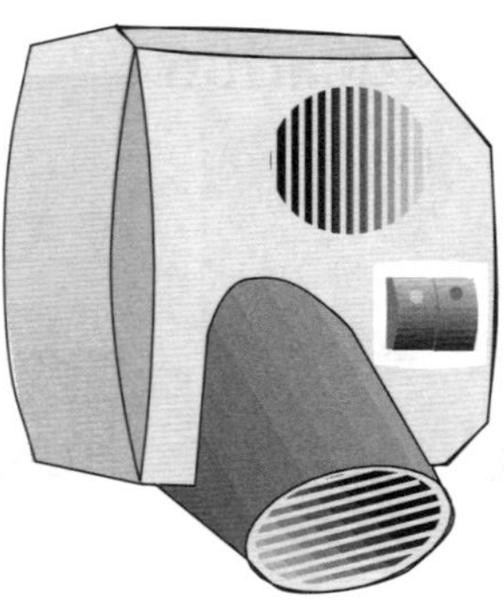

(a) A hand dryer has a current of 10 A flowing through it when it is working normally.

(i) Should the dryer have a 3 A, 5 A, 9 A or 13 A fuse in the plug?

....

(1 mark)

(ii) If a current much higher than normal flows in the dryer, what will happen to the wire in the fuse?

....

....

....

....

(2 marks)

(b) This equation links the energy transferred with the power and the time:

ENERGY TRANSFERRED (in kWh) = **POWER** (in kW) × **TIME** (in h)

(i) The hand dryer is accidentally left on for 3 hours. The power of the hand dryer is 2.4 kW. Work out how much energy is transferred (in kWh) during this time.

....

....

....

(2 marks)

(ii) How much will it cost to run the dryer for 3 hours, if 1 kWh of electricity costs 9p?

....

....

(1 mark)

© CGP 2002

2 This is a graph showing the braking distances for a van travelling at various speeds.

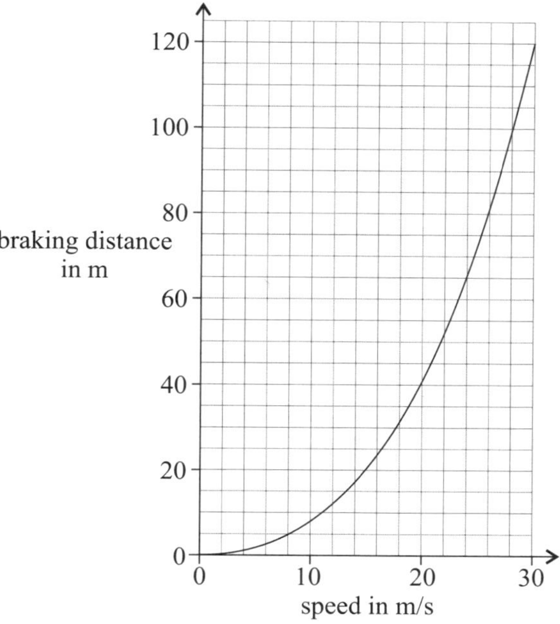

(a) The van driver thinks that if he doubles his speed, he should double the braking distance. Explain whether he is right or not, using the graph to help.

....

....

....
(2 marks)

(b) What would the braking distance be for this van if the brakes were applied when it was travelling at 16 m/s?

....
(2 marks)

(c) The van is driving at 30 m/s and a dog runs into the road ahead.
The driver applies the brakes. Why is his **stopping** distance more than 120 m?

....
(1 mark)

(d) The braking distances in the graph above are applicable in fair weather.
How would braking distances be affected if the roads were icy? Explain your answer.

....

....
(2 marks)

Leave blank

© CGP 2002

Leave blank

3 Shelley sledges down a hill. Here is a graph showing her speed as she travels down the hill.

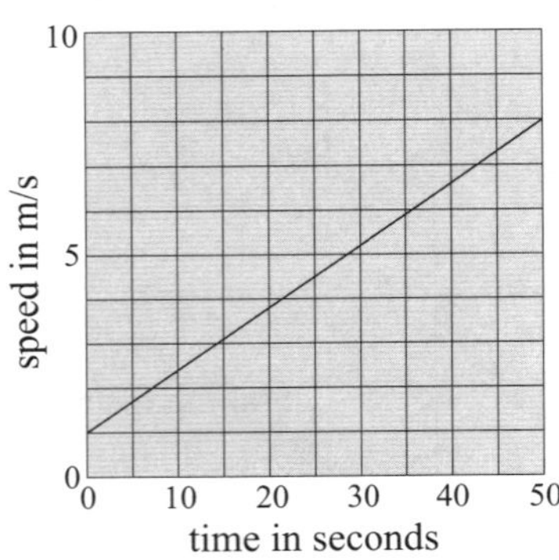

(a) (i) The sledge starts travelling at 1 m/s.
By the time it reaches the bottom of the hill, it is travelling at 8 m/s.

What is the acceleration (in m/s²)?

....

....
(2 marks)

(ii) Describe how the **kinetic** and **potential** energy of the sledge change as it travels down the hill.

....

....

....

....
(2 marks)

(b) When Shelley reaches the bottom of the hill, the sledge continues to travel along the flat for a while before it stops. Here is a diagram showing information about the sledge.

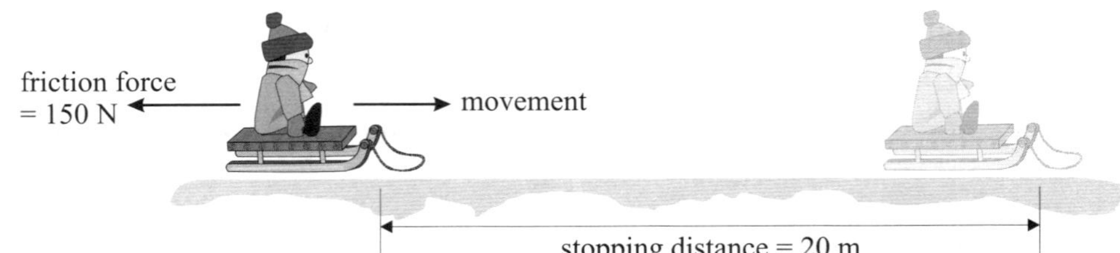

Use the information in the diagram to calculate the work done to stop the sledge.

....

....
(3 marks)

© CGP 2002

(c) Shelley builds an electric winch to get her sledge up to the top of the hill.

The electrical energy supplied to the winch is 6000 J.
The potential energy gained by the sledge is 3000 J.

Calculate the efficiency of the winch.

....

....

....
(3 marks)

Leave blank

4 This diagram shows the electromagnetic spectrum.

radio waves	microwaves	infra-red	visible light	ultraviolet	x-rays	gamma rays

(a) (i) Name one type of radiation which has a lower frequency than infra-red radiation.

....
(1 mark)

(ii) Name a type of radiation which has a shorter wavelength than visible light.

....
(1 mark)

(b) Which type of electromagnetic radiation is used:

(i) in night-vision equipment,

....
(1 mark)

(ii) for treating cancer.

....
(1 mark)

(c) Name two kinds of electromagnetic radiation which can cause cancer.

1. ....

2. ....
(2 marks)

© CGP 2002

5 Here is an electromagnet.

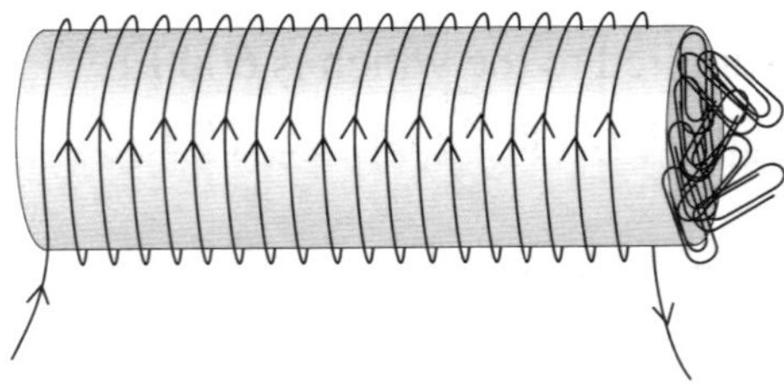

(a) What will happen when the current is switched off?

....

....
(2 marks)

(b) This is the circuit for a resettable fuse.

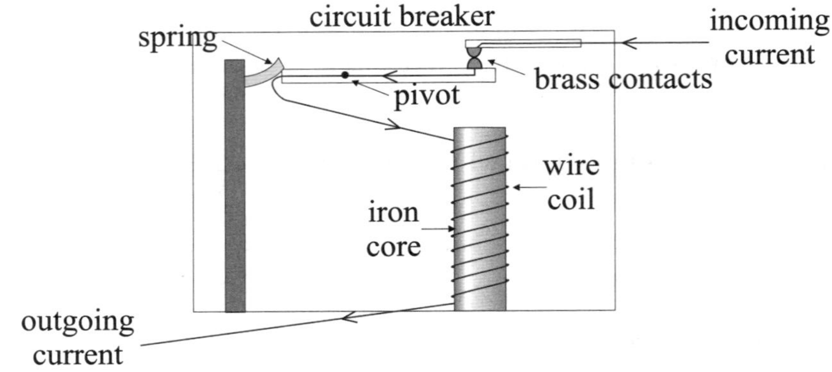

(i) The fuse stops the current flowing when the current in the wire becomes too great. Describe how it works.

....

....

....
(3 marks)

(ii) Why is it called a "resettable fuse"?

....
(1 mark)

Leave blank

© CGP 2002

Leave blank

6 Mildred has made a see-saw for her garden.

(a) When Mildred and Jane sit 1.5 m from the middle, the see-saw will not balance.

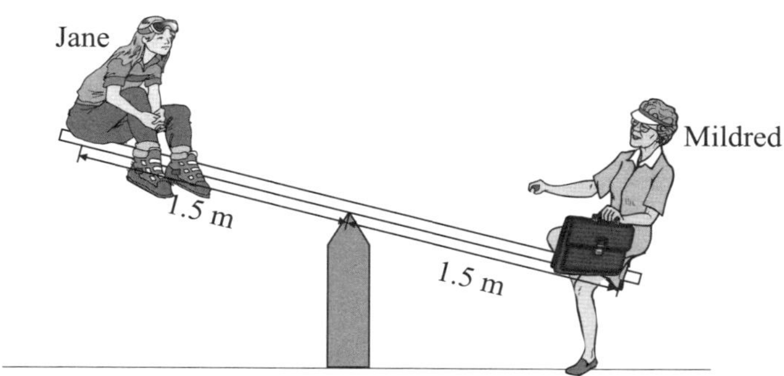

Explain, in terms of the turning force, how you can tell Mildred is heavier than Jane.

....

....

(2 marks)

(b) Mildred moves so that she is sitting just 1.0 m from the middle. The see-saw balances.

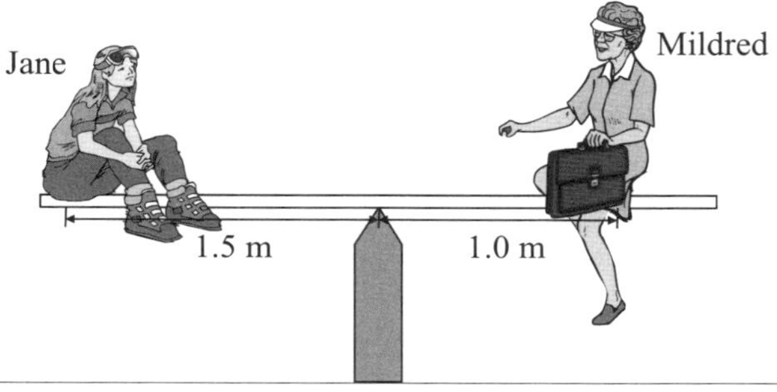

Mildred weighs 700 N, including her briefcase.

Calculate Jane's weight. You will need the following formulas:

moment of a force = force × perpendicular distance from pivot When system is at rest: **sum of clockwise moments = sum of anticlockwise moments**

....

....

....

(3 marks)

Question continued overleaf.

© CGP 2002

(c) Jane does an experiment at school.

Leave blank

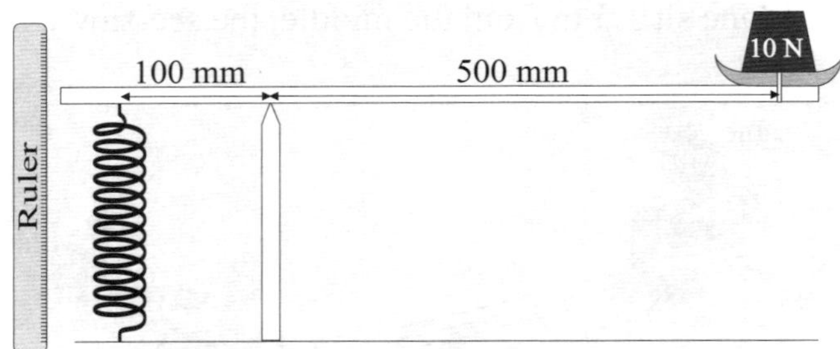

She measures the length of the spring with different weights on the pan. With no weights on, the spring is 300 mm long. This graph shows her results:

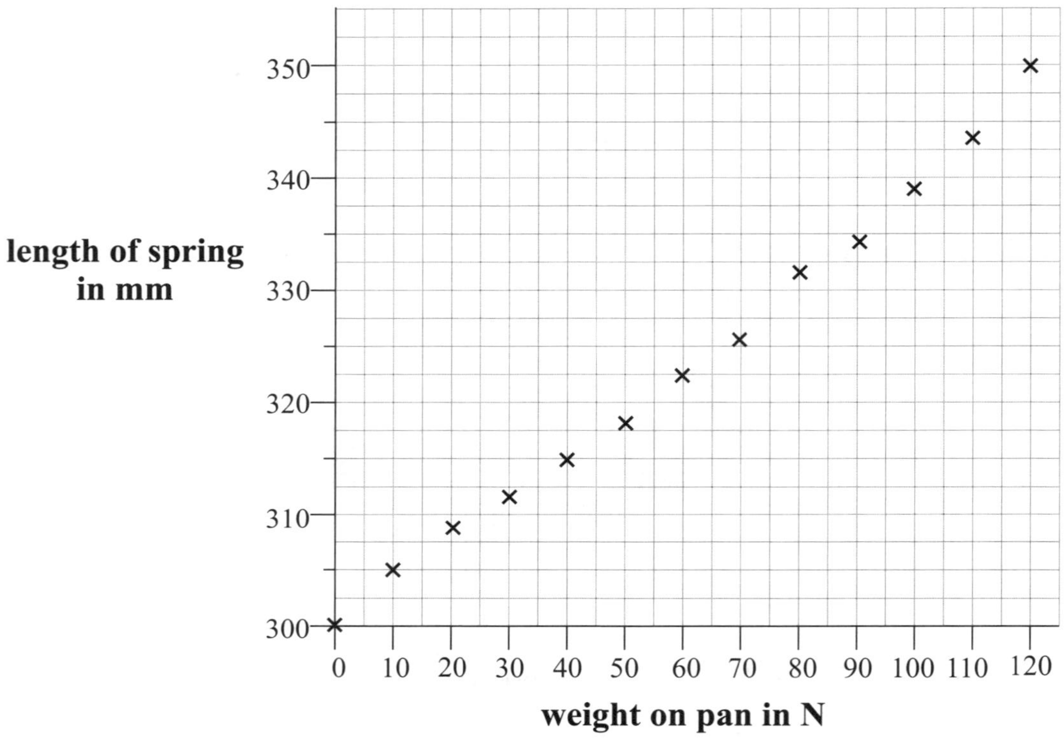

(i) Finish the graph by drawing the best straight line through the points.

(1 mark)

(ii) What would the **extension** of the spring be if a load of 25 N was put in the pan? (Show clearly on the graph how you obtained your answer.)

....

....

(2 marks)

© CGP 2002

(iii) The bar is level when the load applied is 10 N.

Leave blank

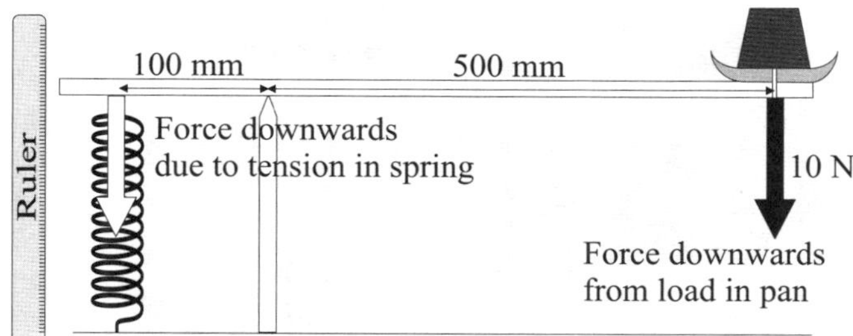

What is the tension force in the spring (in N) when the load is on the pan?
You will need these formulas again:

moment of a force = force × perpendicular distance from pivot When system is at rest: **sum of clockwise moments = sum of anticlockwise moments**

....

....

....

....

(3 marks)

Turn over for next question.

© CGP 2002

Leave blank

7 Mr Vines has built a train set.

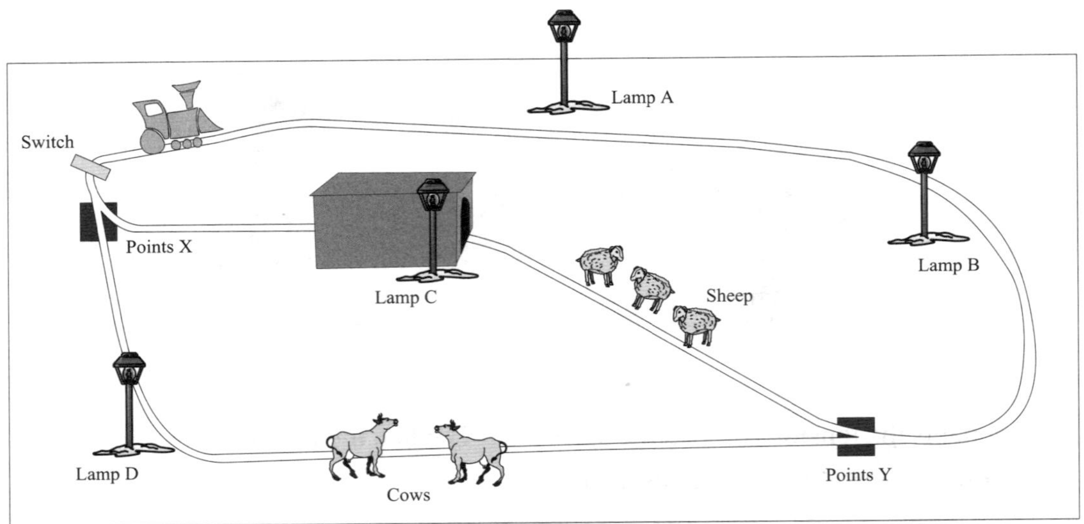

Both sets of points change each time the train runs over a switch under part of the track. The same switch controls which lamps are lit. The circuit is shown below.

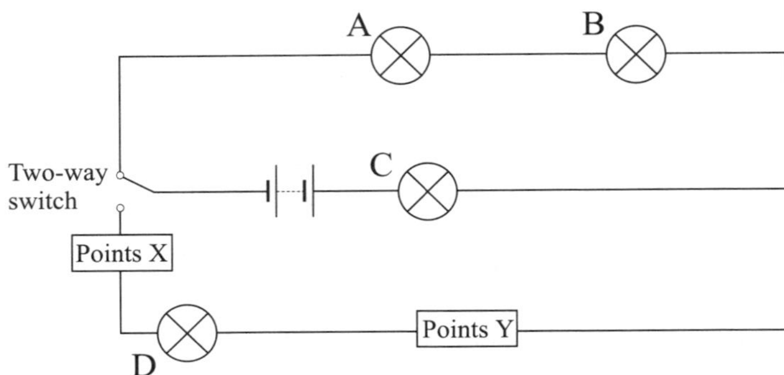

(a) When the points are switched off (ie there is no power supplied to them), the train will take the route past the sheep. Both sets of points change when they are switched on so that the train takes the route past the cows.

Complete this table to show the status of each component when the switch is up or down.

Switch up/down	Points X on/off	Points Y on/off	Lamp A on/off	Lamp B on/off	Lamp C on/off	Lamp D on/off
up						
down						

(6 marks)

© CGP 2002

Leave blank

(b) Mr Vines puts three ammeters into the circuit to measure the current at different points.

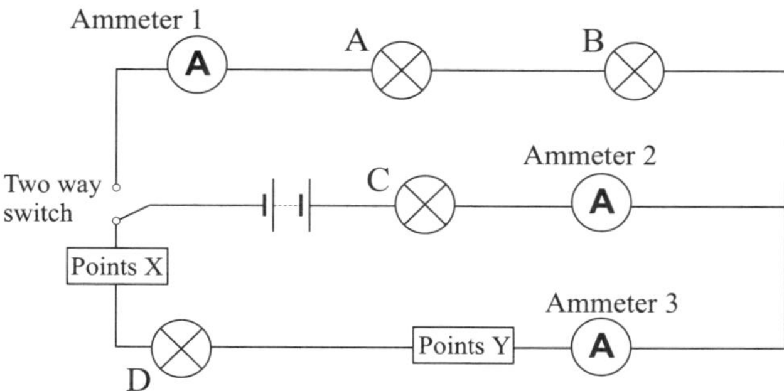

Ammeter 2 is showing a current of 3 A.

(i) What is the reading on ammeter 3?

....

(1 mark)

(ii) What is the reading on ammeter 1?

....

(1 mark)

(c) Mr Vines now puts some voltmeters into the circuit to measure the potential difference across different components. The wires in the circuit have negligible resistance.

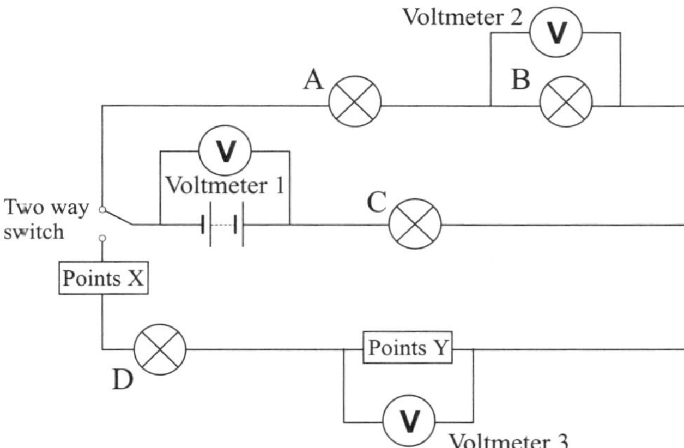

Voltmeter 1 is showing a potential difference of 6 V across the battery.

(i) If each lamp has the same resistance, what is the reading at voltmeter 2?

....

....

(2 marks)

(ii) What is the reading at voltmeter 3?

....

(1 mark)

© CGP 2002

Leave blank

8 Gamma rays are high-energy electromagnetic rays. They can penetrate skin, paper and thin aluminium. Gamma rays cannot penetrate thick lead, so thick lead containers are often used to contain substances which emit gamma rays.

(a) Alpha and beta particles are two other forms of radiation. For each particle, state what it is, and give an example of the kind of material necessary to stop most of the particles.

Your answers should clearly reflect the different penetrating powers of the two types of radiation.

Alpha particle: ....

....

Beta particle: ....

....

(4 marks)

(b) Iodine-131 gives out beta particles and gamma rays, and it has a half-life of 8 days. It is used as a tracer in the body to examine thyroid glands.

(i) For tracers in the body, radioactive substances which emit alpha particles must not be used. Explain why.

....

....

(2 marks)

(ii) Many radioactive isotopes, including iodine-131, do not occur naturally, and must be made artificially. Why is a radioactive substance with a half-life of 8 days used for medical tracers, instead of a substance with a half-life of 8 minutes or 8 years?

....

....

....

(3 marks)

(c) Carbon-14 is a radioactive isotope with a half-life of 5600 years.
An archaeologist finds a bone which she takes to be carbon-dated. It is discovered that the carbon-14 in the bone has decayed to one eighth of its original amount.
The bone is found to be 16 800 years old. Explain how the calculation was made.

....

....

(2 marks)

© CGP 2002

9 This diagram shows an electrolysis experiment to coat a coin with a layer of silver.

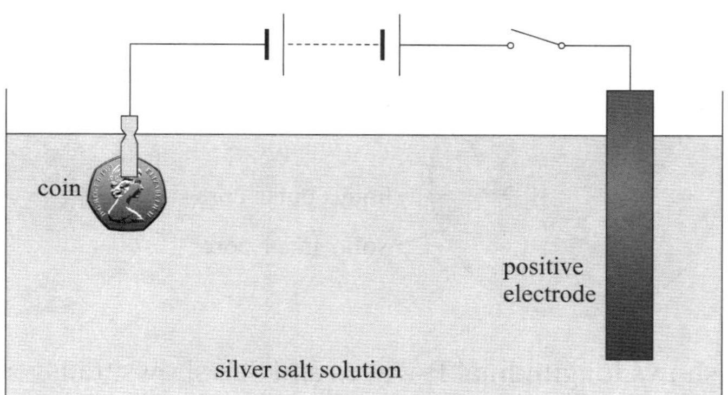

When the switch is closed, electric charge flows around the circuit.

(a) (i) Explain how the charge is carried through the **wires** in the circuit.

....

....

....

(1 mark)

(ii) Explain how the charge is carried through the **silver salt solution**.

....

....

....

....

(3 marks)

(b) The switch is closed.
A charge of 800 coulombs flows around the circuit with a current of 10 A.
How long does this take?

....

....

(2 marks)

© CGP 2002

Leave blank

10 When there is an earthquake, seismographs all over the world detect the seismic waves. These diagrams show how seismic waves travel through the Earth.

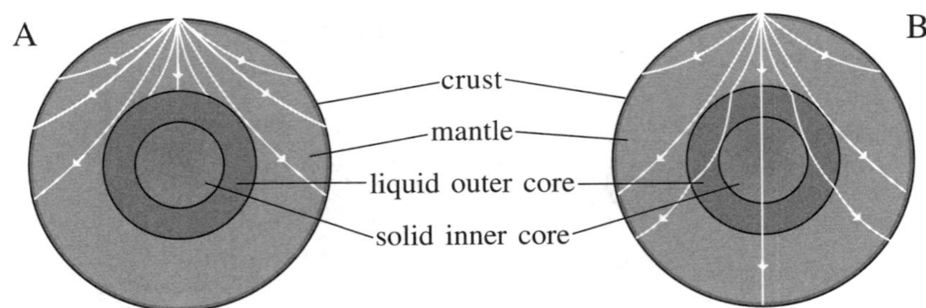

(a) One diagram shows longitudinal P-waves and one shows transverse S-waves. Name the type of waves shown in:

(i) diagram A,

....

(ii) diagram B.

....

(2 marks)

(b) (i) The waves travelling through the solid crust and mantle have a curved path. Explain why.

....

....

....

(3 marks)

(ii) In diagram B, the waves suddenly change direction when they reach the outer core. Explain why.

....

....

(2 marks)

© CGP 2002

11 Optical fibres use total internal reflection to carry electromagnetic waves.

(a) This diagram shows the path of an electromagnetic wave along an optical fibre.

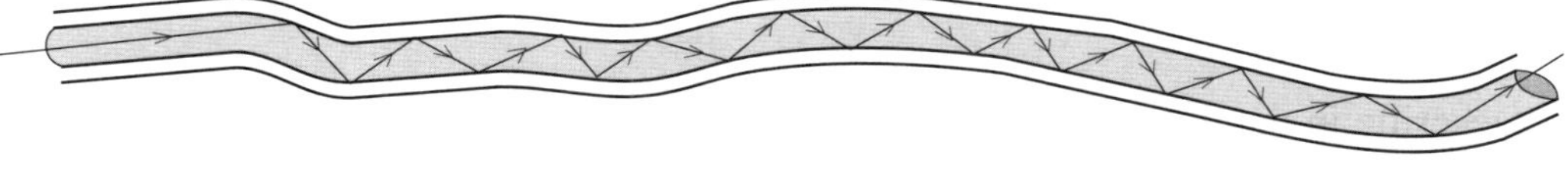

Explain how the electromagnetic wave moves along the fibre.

....

....

....
(2 marks)

(b) Optical fibres are used in telecommunications to carry information over long distances. Give **two** advantages of using optical fibres over using electrical signals in wires.

1. ....

....

2. ....

....
(2 marks)

12 Long wave radio is received more easily in hilly areas than shorter wave FM radio. The diagrams below show FM radio and long wave radio being transmitted. Complete the wave diagrams of the different radio waves and use the diagrams to explain:

(a) how the house in the diagram receives long wave radio,

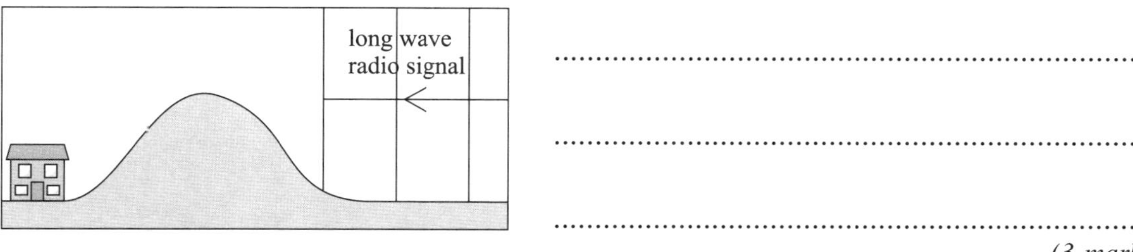

....

....

....
(3 marks)

(b) why the house in the diagram does not receive FM radio.

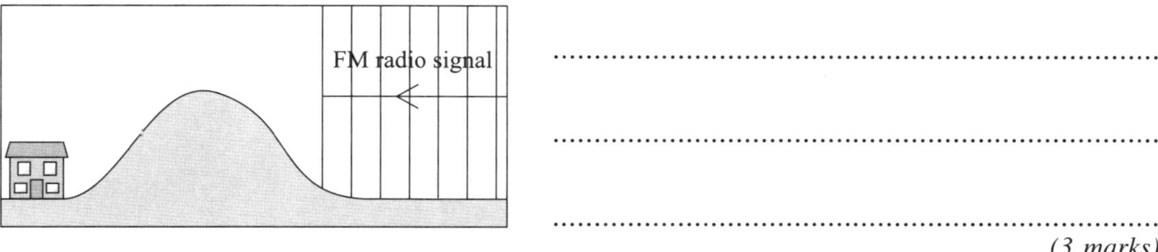

....

....

....
(3 marks)

Leave blank

© CGP 2002

Leave blank

13 (a) Write down the name of the force that keeps planets in orbit around the Sun.

....
(1 mark)

(b) Describe in detail the orbit of a planet in our solar system.

....

....

....
(3 marks)

(c) Comets are only rarely seen from Earth.
Explain why this is by describing a comet's orbit.

....

....

....
(3 marks)

(d) There are many objects orbiting the Earth, including a number of artificial satellites.
Give **two** different uses of artificial satellites.

1. ....

....

2. ....

....
(2 marks)

PHP4U

© CGP 2002

General Certificate of Secondary Education

GCSE
Science: Double Award (Coordinated)
Paper 3 – *Higher Tier*

Centre name					
Centre number					
Candidate number					

Physics

Surname
Other names
Candidate signature

Time allowed: 1 hour 30 minutes.

Instructions to candidates
- Write your name and other details in the spaces provided above.
- Answer **all** questions in the spaces provided.
- Do all rough work on the paper.
- Write your answers in black or blue ink or ball-point pen.

Information for candidates
- The marks available are given in brackets at the end of each question or part-question.
- Marks will not be deducted for incorrect answers.
- In calculations show clearly how you work out your answers.
- State the units in all your answers.
- There are 13 questions in this paper. There are no blank pages.

Advice to candidates
- Work steadily through the paper.
- Don't spend too long on one question.
- If you have time at the end, go back and check your answers.

For examiner's use

Q	Attempt Nº 1	2	3	Q	Attempt Nº 1	2	3
1				8			
2				9			
3				10			
4				11			
5				12			
6				13			
7							
				Total 100			

© CGP 2002

Leave blank

1 (a) The heating element of a kettle is near the bottom.

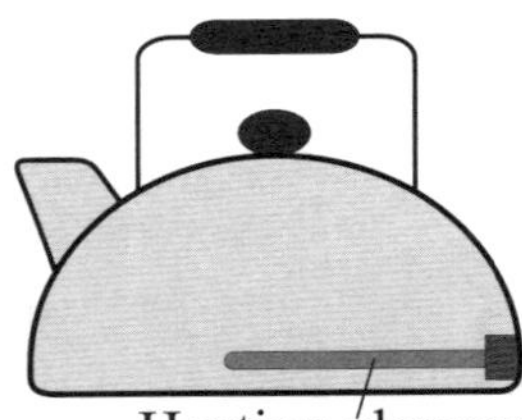

(i) Explain why the heating element is not put near the top of the kettle.

....

....
(2 marks)

(ii) If this kettle takes 3.5 minutes to boil some water, using 420 000 joules of electrical energy in the process, use the formula

energy = power × time

to calculate the power (in watts) of the kettle.

....

....

....

Power: .... *watts*
(3 marks)

(b) A hot liquid in an open container loses heat by evaporation.
Explain how this evaporation leads to a decrease in the temperature of the liquid.
Use ideas about the motion of particles in your answer.

....

....

....

....
(3 marks)

© CGP 2002

Leave blank

(c) Bill decides he needs a new convection heater for his living room.
Explain how a convection heater heats a room.

....

....

....
(3 marks)

(d) The salesman tells Bill that a convection heater with a fan to blow air over the heating element is more efficient. Why is this? Explain your answer fully.

....

....
(1 mark)

Turn over for next question.

© CGP 2002

Leave blank

2 (a) The following list shows the electromagnetic spectrum in order.
Fill in the blanks.

radio waves

....

infrared
visible light

....

X-rays
gamma rays

(2 marks)

(b) Which part of the spectrum has the **shortest** wavelength?

....

(1 mark)

(c) All electromagnetic waves travel at the same speed.
If one wave has a shorter wavelength than another, what can you say about its **frequency**?

....

(1 mark)

(d) The nature of sound waves is very different to that of electromagnetic waves.
Briefly describe this difference.

....

....

(2 marks)

© CGP 2002

Leave blank

3 This question is about radioactive isotopes.

(a) (i) At the start of an experiment, the activity count of a radioactive isotope was 1600 counts per second, after allowing for background radiation.
After an hour, the count was 200 counts per second.
Calculate the half-life of the isotope in minutes.

....

....

Time: .... *minutes*
(2 marks)

(ii) What will the activity count be after one more hour?

....

....

Activity count: .... *counts per second*
(2 marks)

(b) Emissions from the three forms of radioactive decay can be stopped by different kinds of barrier. Give an example of the sort of material necessary to block the following.

Your answers should clearly reflect the different penetrating powers of the three types of radiation.

(i) alpha particles,

....

(ii) beta particles,

....

(iii) gamma rays.

....
(3 marks)

© CGP 2002

Leave blank

4 Ingo takes part in the sport of speed skiing. The aim is to go as fast as possible down a straight but very steep course.
Ingo starts by stepping out of a special hut at the top of the course. After about 25 seconds, he reaches his terminal speed.

(a) (i) What force causes him to accelerate down the hill?

....

....
(1 mark)

(ii) What happens when he reaches terminal speed and why?

....

....
(2 marks)

(iii) Explain one method Ingo could use to increase his terminal speed.

....

....
(1 mark)

(b) A graph of his motion is shown below.

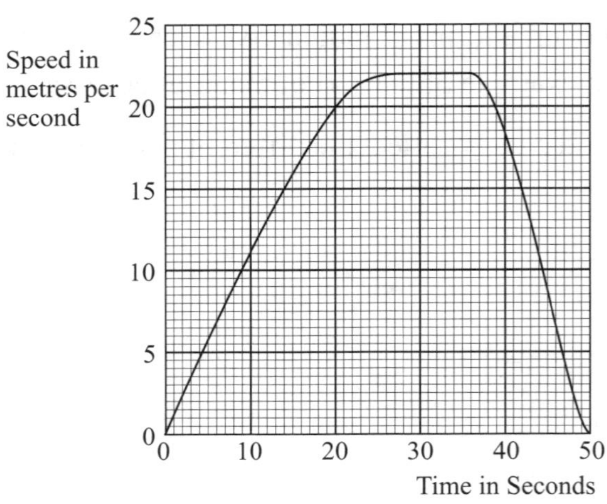

(i) How far did Ingo travel between 28 and 35 seconds?

....

....
(2 marks)

(ii) Describe Ingo's motion after 35 seconds.

....

....
(1 mark)

© CGP 2002

5 In the well-known sport of bungee running, an elastic cord is tied to a runner's waist. The runner then has to run as far as possible before being pulled back by the cord. The diagram below shows a bungee runner at different stages of his run. The horizontal forces A, B, C, D and E act on him at different times.

Leave blank

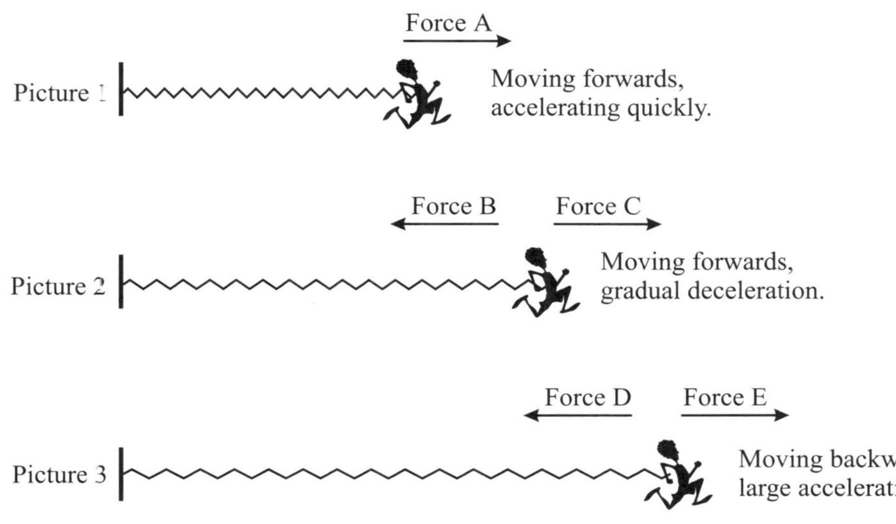

(a) By considering the horizontal forces A, B, C, D and E, describe how the motion is caused in

(i) Picture 1,

....

(ii) Picture 2,

....

(iii) Picture 3.

....

(3 marks)

(b) At a time somewhere between pictures 1 and 2, the runner is neither accelerating nor decelerating. What can you say about the forces acting on the runner at this point?

....

....

(1 mark)

© CGP 2002

Leave blank

6 The diagram shows a teapot on a table.

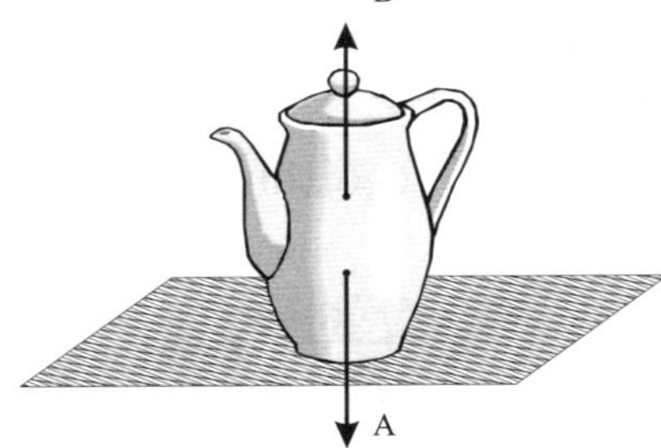

(a) What causes forces A and B?

A: ....

B: ....

(2 marks)

(b) (i) How can you tell that force A is equal in size to force B?

....

....

(1 mark)

(ii) What can you say about the directions of forces A and B?

....

(1 mark)

(c) Ivan pours two cups of tea from the teapot, then puts the teapot back on the table.

What happens to each of the forces A and B? Choose from the phrases below.

becomes bigger **stays the same** **becomes smaller**

A: ....

B: ....

(2 marks)

© CGP 2002

Leave blank

7 A new power station is needed, as an existing one is coming to the end of its useful life. The current plan is to build either a coal-fired power station or a nuclear power station.

(a) Describe the advantages **and** disadvantages of each of these methods of generating electricity.

....

....

....

....

....

....

....

....

....

....

(6 marks)

(b) A local environmental group has suggested that instead of using coal or nuclear power, a renewable source of energy should be used.

(i) Give an example of a **renewable** energy resource.

....

(1 mark)

(ii) Give an advantage **and** a disadvantage of this energy source.

Advantage: ....

Disadvantage: ....

(2 marks)

© CGP 2002

Leave blank

8 The handlebar grips on James' motorbike are heated. When the heaters in the grips are switched on one cold December morning, 18 000 joules of electrical energy are transferred to the heating elements, and their temperature is raised by 30 °C.

Heating elements inside grips

The combined mass (m) of the metal elements in the handlebar grips is 0.1 kg.
The heating elements are made of a metal with specific heat capacity $S = 1100$ J/kg°C.

The energy (E) required to raise the temperature of a mass m of metal by x °C, where the metal has specific heat capacity S is given by:

$$E = m \times S \times x$$

(a) Calculate the energy required to raise the temperature of the heating elements by 30 °C.

....

....

Energy: .... *joules*

(2 marks)

(b) Use the formula below to calculate the efficiency of the handlebar grips.

$$\text{efficiency} = \frac{\text{useful energy transferred}}{\text{total energy transferred}} \times 100\ \%$$

....

....

Efficiency: .... %

(2 marks)

(c) How does turning the heaters on affect the kinetic energy of the particles in the heating elements?

....

....

(1 mark)

© CGP 2002

Leave blank

(d) The heating elements could be controlled by sensors in the handlebar grips that respond to changes in temperature. Explain why a thermistor might be used in this kind of sensor.

....

....

....

(2 marks)

(e) The electrical equipment on a motorbike is connected in parallel rather than in series. Explain why.

....

....

....

(2 marks)

9 The diagram below shows the main parts of a power station.

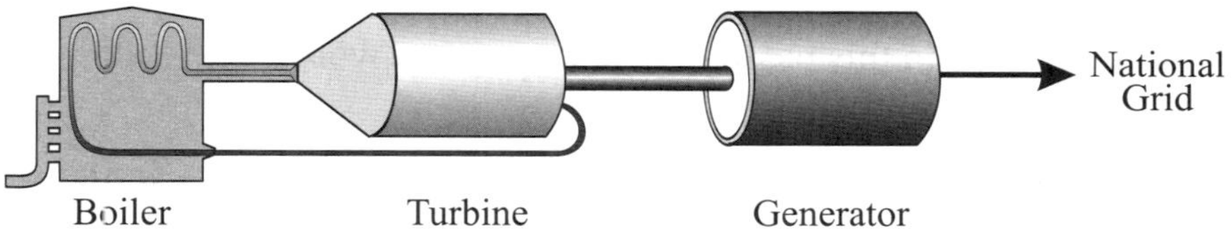

Boiler Turbine Generator National Grid

Using the diagram to help, explain how electricity is generated.

....

....

....

....

....

....

(4 marks)

© CGP 2002

Leave blank

10 Below is a diagram of the Solar System. It is not to scale.

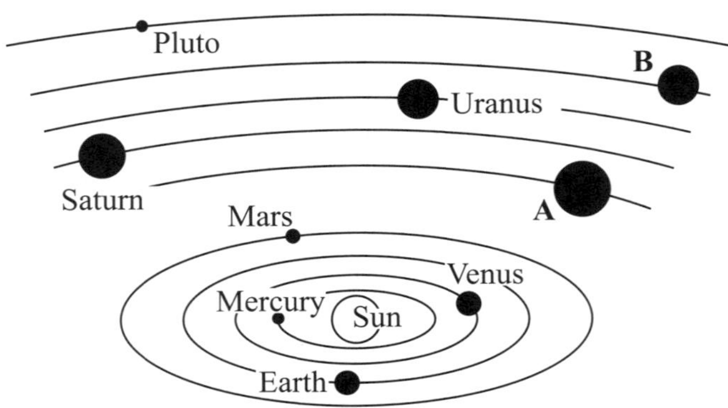

(a) The names of the planets marked A and B are missing. What are their names?

A: ....

B: ....

(2 marks)

(b) It is sometimes possible to see Mars and Venus in the sky with the naked eye, although they do not give out any light. Explain how this is possible.

....

....

(1 mark)

(c) (i) Fill in the gaps in this passage with the correct words.
Choose from the alternatives below.

eight **weaker** **two** **stronger** **four** **three**

Uranus is approximately twice as far from the Sun as Saturn.

This means the Sun's gravity at Uranus is .... times .... than the Sun's gravity at Saturn.

(2 marks)

(ii) Which of the planets travels fastest through space? Explain your answer.

....

....

(2 marks)

© CGP 2002

Leave blank

11 Look at the following picture of a transformer.

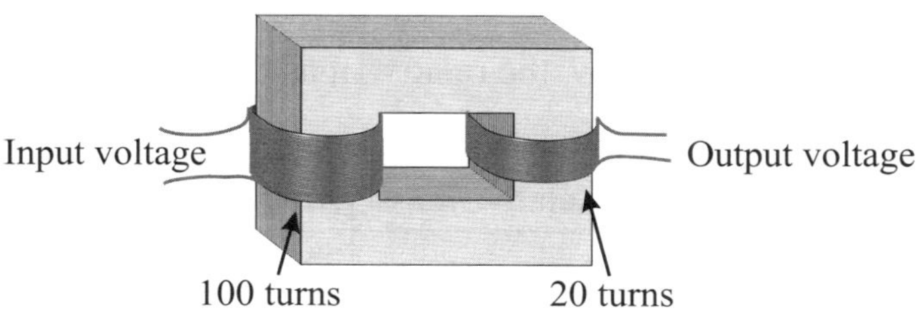

(a) Is this an example of a **step-up** transformer or a **step-down** transformer?
How can you tell?

....

(2 marks)

A **100 V d.c.** input voltage is connected to the transformer.

(b) Explain what happens.

....

....

(1 mark)

(c) High voltages are used to transmit electricity over large distances. Explain why.

....

....

(2 marks)

Turn over for next question.

© CGP 2002

Leave blank

12 Nowadays we believe that atoms are made up of protons, neutrons and electrons.

(a) Complete the table below to show the relative mass and charge of these three types of particle.

Particle	Relative mass	Relative charge
proton		
neutron		
electron	$^{1}/_{2000}$	–1

(2 marks)

(b) (i) Which **two** diagrams of atomic nucleii below show different isotopes of the same element?

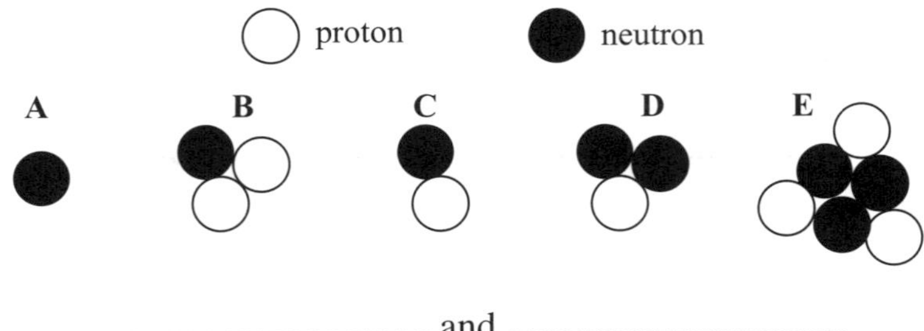

.... and ....

(1 mark)

(ii) Explain what is meant by the phrase 'different isotopes of the same element'.

....

....

....

(2 marks)

Beta particles are ejected from the nucleus when certain radioactive elements decay.

(c) Describe a beta particle, and the changes that occur in the nucleus of an atom when a beta particle is emitted.

....

....

....

....

(3 marks)

© CGP 2002

(d) The diagram shows a device used to control the thickness of paper. A radiation detector is used to detect how much radiation from the radiation source penetrates the paper. If too much radiation penetrates, the paper is too thin and so the hydraulic control is adjusted. If too little radiation penetrates, the paper is too thick and the opposite adjustment is made.

Leave blank

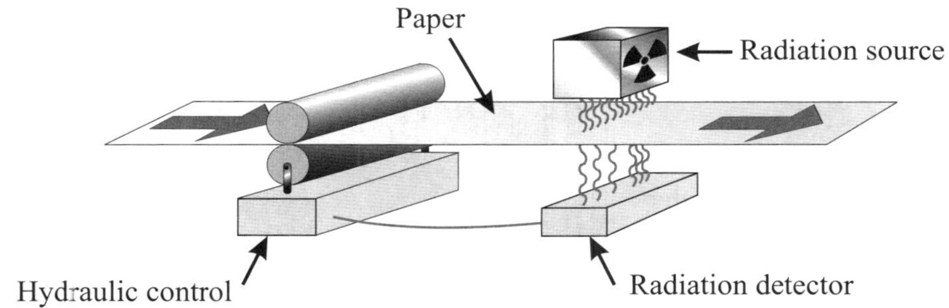

(i) Would **alpha particles**, **beta particles** or **gamma rays** be the most suitable type of radiation to use in this device? Explain your answer.

....

....

(2 marks)

(ii) A radiation source with a very short half-life would be unsuitable here. Explain why.

....

....

(2 marks)

(iii) A similar device can be used to ensure sheets of metal are the correct thickness. Why would you need to use a different source of radiation in this case? What type of radiation source would you recommend for this new type of machine?

....

....

....

(2 marks)

© CGP 2002

13 This question is about television and radio signals.

Leave blank

(a) Describe one **similarity** and one **difference** between radio waves and visible light.

Similarity: ....

Difference: ....

(2 marks)

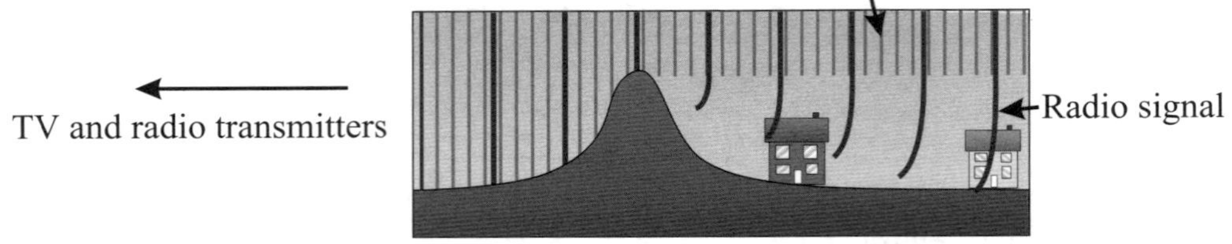

TV signal

TV and radio transmitters

Radio signal

(b) The diagram above shows two houses situated near a hill. Explain why they are able to receive long-wave radio signals, but cannot pick up shorter wavelength TV signals.

....

....

(2 marks)

(c) A typical radio wave might have a wavelength of 300 metres and travel at a speed of 300 000 000 m/s.

(i) Draw a diagram to show clearly what is meant by the phrase 'a wavelength of 300 metres'.

(2 marks)

(ii) Calculate the frequency of a radio wave whose wavelength is 300 metres.

....

Frequency: .... *Hz*

(2 marks)

(d) Sound also travels by means of waves. Explain briefly how sound waves are different to radio waves.

....

....

(2 marks)

PHP4U

© CGP 2002

Coordination Group Publications

GCSE

Double Science (Coordinated)

Answer Book

Practice Exam Papers

Higher Tier

These practice papers won't make you better at science

... but they will show you what you **can** do, and what you **can't** do.

These are GCSE papers, just like you'll get in your exams — so they'll tell you what you need to **work at** if you want to do **better** on the day.

Do an exam, **mark it** and look at what you **got wrong**.
That's the stuff you need to learn.

Go away, **learn** those tricky bits, then **do the same exam again**. If you're **still** getting questions wrong, you'll have to do even **more practice** and **keep testing** yourself until you keep getting **all** the questions right.

It doesn't sound like a lot of **fun**, but it **really will help**.

The three big ways to improve your score

1) **Answer all these exams**
These practice papers contain all the types of question that have come up year after year in GCSE exams. If you can do all these, you should be able to do all the questions in your exams.

2) **Keep practising the things you get wrong**
The whole point of a practice exam is to find out what you don't know*. So every time you get a question wrong, revise that subject then have another crack at it.
*Use the mark scheme in this booklet to help you see where you dropped your marks.

3) **Don't throw away easy marks**
Always answer the question the way it's asked — if it asks for units, use the right ones. Always double-check your answer and don't make silly mistakes — obvious really.

Remember: the fewer marks you lose, the more marks you get.

© 2002 CGP

Working out your Grade

- In the real exam, you'll do one Physics, one Biology and one Chemistry paper. This pack contains practice Physics papers.
- Use the answers and mark scheme to mark each paper. The marks are all out of 100, so they're already percentages.
- If you've done papers in Chemistry and Biology as well, find your average percentage for the whole exam.
- Look up your mark in this table to see what grade you got. If you're borderline, don't push yourself up a grade — the real examiners won't.

Average %	85+	74 – 84	61 – 73	47 – 60	37 – 46	30 – 36	under 30
Grade	**A***	**A**	**B**	**C**	**D**	**E**	**U**

Stick your marks in here so you can see how you're doing

(We've included space for your Chemistry and Biology results as well.)

		Physics	Chemistry	Biology	Average %	Grade
EXAM 1	First go					
	Second go					
	Third go					
EXAM 2	First go					
	Second go					
	Third go					
EXAM 3	First go					
	Second go					
	Third go					

Important!

Any grade you get on one of these practice papers is **no guarantee** of getting that in the real exam — **but** it's a pretty good guide.

© 2002 CGP

Physics Paper 1 - Higher Tier

1 **(a)** **(i)** Thinking distance ***increases*** with speed. [1 mark]

(ii) You need the formula: distance = speed × time. So if the speed increases and the time stays the same, the distance travelled also increases.
[1 mark for stating the formula **or** giving a verbal explanation of it]

(b) Any two of the following: tiredness, alcohol, drugs, illness, age [1 mark each]
(If you put visibility, you don't get a mark. It affects the time taken for the driver to see the danger, but not the driver's reaction time.)

(c) Distance increases [1 mark] because kinetic energy (which has to be converted by brakes to heat etc.) is proportional to mass. More kinetic energy means more time, and therefore a longer distance, to stop [1 mark]

Stopping distance is a favourite topic for the exams. Make sure you know what affects thinking distance and braking distance... and that you know braking distance is proportional to speed² (because of the v^2 bit in kinetic energy (=½ mv^2)).

(d) **(i)** ***14 m/s*** [1 mark for an answer between 13.5 m/s and 14.5 m/s.]

(ii) deceleration = change in velocity ÷ time taken = 45 ÷ 7.2 = ***6.25m/s²***
[1 mark for 45 ÷ 7.2 or equivalent if you used your answer to part (i) and 5 s, 1 mark for answer between ***6.1*** and ***6.3 m/s²*** (if you used your answer to part (i))]

(iii) F = ma = 600 × 6.25 = ***3750 N***
[1 mark for 600 × answer to part (ii), 1 mark for correct answer]

(iv) To find the distance, you just work out the area under the line on the graph.
distance = area under graph = ½ × time × change in velocity
(using formula for area of triangle)
So, distance = ½ × 7.2 × 45 = ***162 m***
{Or you could use the formula: distance travelled = average speed × time taken.
The average speed is given by:

$$\text{average speed} = \frac{\text{start speed} - \text{end speed}}{2} = \frac{45-0}{2} = 22.5 \text{ m/s}$$, so the distance

travelled while slowing down must be 22.5 × 7.2 = 162 m}
[1 mark for working out the area under the graph (or using a suitable alternative method), and 1 mark for the right answer]

---

2 **(a)** **(i)** ***Galaxy*** (lots of stars) [1 mark]

(ii) ***Comet*** (just lumps of rock and ice) [1 mark]

(b) Stars are only visible at night because the Sun scatters too much light through the atmosphere during the day. ***Or*** the light from the stars is drowned out by the light from the Sun during the day. [1 mark for either of these]

(c) Once the star's ***hydrogen/fuel is used up***, the star will ***expand*** and become a ***red giant***. It then ***cools and contracts*** under gravity to become a white dwarf.
[1 mark for each of any 3 of the points in bold]

I know. I'm a star.

© 2002 CGP

3 **(a)** **(i)** ***Negative*** [1 mark]

(ii) Copper ions are positive so they are attracted to, and deposited at, the negative electrode (or cathode). [1 mark for writing "opposites attract" or an equivalent expression, 1 mark for saying why.]

(b) ***Decreasing voltage of d.c. supply.*** [1 mark], ***increasing resistance of R.*** [1 mark].

---

4 **(a)** The tyre rotates the magnet inside a fixed coil. [1 mark]
The magnetic field lines move through the coil. [1 mark]
A current is induced in the coil. [1 mark]

(b) ***Alternating current*** [1 mark]. As the magnet rotates, the magnetic field through the coil changes direction every half-turn. [1 mark]

If there's one thing certain in life it's that you'll get a question on electromagnetism. So if it's a bit shaky now, you'd better get learning before the exam.

(c) Any three of the following [1 mark for each]: Turn the bicycle wheel faster; use a stronger magnet; put more turns on the coil; increase the area of the coil.

---

5 **(a)** **(i)** Work is done against gravity. The force necessary is equal to the weight (in newtons) of the cart (ignoring friction because we're only talking about <u>useful</u> work). The distance travelled is the vertical height raised.
work = force × distance = 6500N × 180 m = 1 170 000 joules = ***1170 kJ***
[1 mark for correct formula, 1 mark for substituting in values; 1 mark for answer]

(ii) power = work done ÷ time taken = 1 170 000 ÷ (4 × 60) = ***4875 watts (or 4.875 kW)***. [1 mark for answer part (i) ÷ time, 1 mark for remembering to convert time to seconds, 1 mark for correct answer]

(b) 1. Extra power is needed to overcome frictional forces in the pulleys, cable and track.
2. Need spare capacity for safety (the cart may not be empty, the weights of individual carts may differ etc.) [1 mark for "friction", 1 mark for other reasonable answer]

---

6 **(a)** The half-life is the time taken for the radiation to reduce by one half. [1 mark]

(b) Measure the amount of carbon-14 in the skeleton [1 mark] and in living/recently deceased bone (eg from a recently deceased animal). Use the half-life to estimate how long it's taken for the reduction to occur. [1 mark]

(c) **(i)** Newly dead material will have 1 atom of carbon-14 in 10 million carbon atoms. After 1 half-life it'll have 1 atom of ^{14}C in 20 million C atoms [1 mark], and after 2 half-lives it'll have 1 atom of ^{14}C in 40 million C atoms [1 mark]. So the skeleton is about ***2 half-lives*** old [1 mark].

(ii) 5600 × 2 = ***11 200 years*** [1 mark]

(d) Stone was never a living material, so it can't be carbon dated. [1 mark]

(e) It is one half-life old, so it was buried ***after*** the original skeleton. [1 mark]

Lonely, shy, carbon rod seeks older person for fun and friendship...

© 2002 CGP

7 **(a)** **(i)**

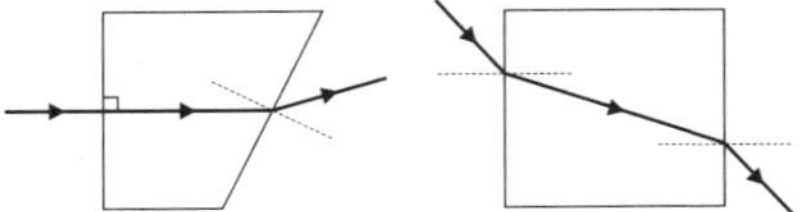

[2 marks — 1 for getting each block right. The angles aren't important, just whether it bends to the right or left or whether it carries straight on.]

(ii)

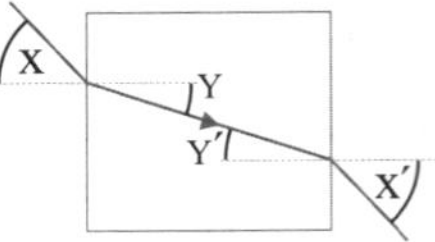

[2 marks — 1 for each pair of angles]

(b) Diagram 1: the angle of incidence (i) > the critical angle so total internal reflection occurs and all the light is reflected to Y [1 mark].
Diagram 2: angle of incidence < critical angle so most of the light is refracted to X and only a small proportion is reflected to Z. [1 mark].
[1 extra mark for 'total internal reflection' or 'critical angle'].

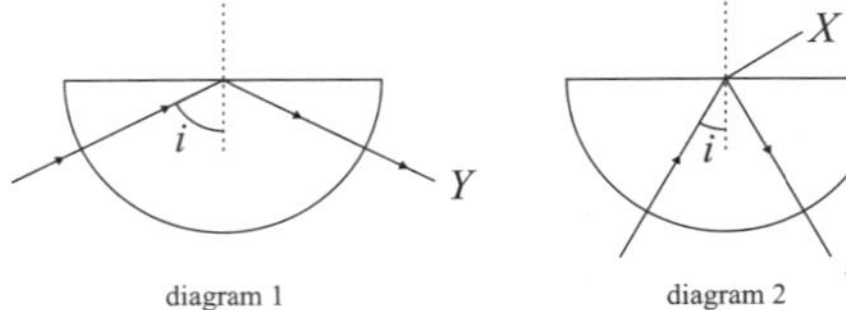

This is a pretty bog-standard question. The only other thing they might ask you for is uses of total internal reflection — binoculars, communications, endoscopes, etc.

---

8 **(a)** **(i)**

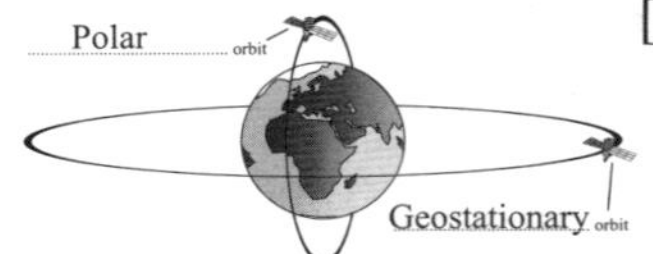

[1 mark for both correct]

(ii) The orbital time is much shorter [1 mark] (just a few hours) — because it's in a lower orbit [1 mark].

(b) 1. Lower orbit so better for photography, spying etc. [1 mark]

2. Scans whole Earth every day as it rotates below path of satellite. [1 mark]

(c) Best suited to: transmission of radio/telephone/TV signals or 'communication',
Also used for: weather monitoring, navigation, astronomy. [1 mark for any of these]

---

9 **(a)** The sweat evaporates. Only the fastest water molecules evaporate, ie the ones with the most energy [1 mark]. That means the average energy of the remaining molecules will be lower [1 mark] — so the body temperature is reduced [1 mark].

(b) Black absorbs a lot more radiant heat, silver tends to reflect it. Adam's bottle absorbs more radiant heat from the Sun than Andrew's. [1 mark for black absorbs / silver reflects, 1 mark for "radiant heat" or "radiation"]

© 2002 CGP

(c) Any two reasonable insulation methods [1 mark for each] with a correct explanation relevant to that method [1 mark for each].

Evaporation's pretty easy to understand — just make sure you can remember all the picky details — they're bound to ask you for them.

---

10 **(a)** **(i)** The vibrations of the medium (air/water etc.) are in the same direction as the direction of travel of the wave (or 'along the line of the wave'). [1 mark]

(ii) Ultrasound is sound of too high a frequency to be heard by the human ear. [1 mark]

(iii) The frequency would remain unchanged (velocity = frequency × wavelength) [1 mark], but the wavelength would increase. [1 mark]

(b) Diffraction is when waves passing through a narrow gap spread out.
With the source as on the diagram, ultrasonic waves can be detected in an arc beyond the metal sheet, not just directly in front of the gap. See diagram below.

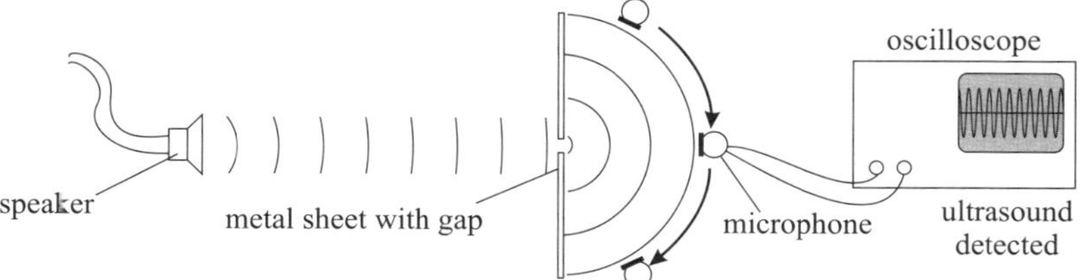

[1 mark for definition of diffraction, 2 marks for clear explanation or diagram]

(c) Ultrasound is relatively harmless but X-rays are harmful to life.
X-rays are unlikely to be reflected.
Metal waste on the sea floor and metal on the boat may alter readings.
X-rays are harder to generate — needs a greater power source.
X-rays are more dangerous for people carrying out the survey.
X-ray equipment is more bulky/expensive.
[1 mark for each valid point, up to a maximum of 3.]

Who said science wasn't useful... only the other day, I was up at Loch Ness. If only I'd had that ultrasound equipment...

---

11 Glass smashes, releasing spring which closes the switch ⟶ completes the circuit ⟶ electromagnet on ⟶ attracts iron bar ⟶ rings bell and breaks circuit ⟶ electromagnet off, spring pulls iron bar back ⟶ complete circuit again ⟶ whole thing repeats, ringing bell continuously. Nice.
[1 mark for switch, 1 mark for electromagnet attracting bar, 1 mark for bell ringing and breaking circuit, 1 mark for repeat]

---

12 **(a)**

Particle	Relative mass	Relative charge
proton	1	+1
neutron	***1***	***0***
electron	***1/2000*** *(approx)*	***–1***

[1 mark for each particle completely correct]

© 2002 CGP

(b) ***D and E*** [1 mark]. The nuclei have the same number of protons (but different numbers of neutrons). (It's the same number of protons in the nucleus that makes them isotopes of the same element.) [1 mark for saying what an isotope is.]

(c) (i) Nuclear radiation can cause ***ionisation*** of molecules in body cells, which in turn can ***damage or destroy the cells***. This can lead to ***cancer*** or ***radiation sickness***. Nuclear radiation can also cause ***mutations in DNA***.
[1 mark for any of the ideas in bold]

(ii) Any two from:
Wearing (lead-lined) protective suits; keeping behind lead/concrete barriers; using remote controlled robot arms to handle sources of radiation; regularly checking exposure to radiation. [1 mark for each of 2 precautions]

Radioactivity... dum de dum... nope... nothing useful to say.

13 (a) (i) The speed ***increases*** (i.e. the mintball ***accelerates***). [1 mark]

(ii) This is because ***Force P*** (weight due to gravity) ***is greater than Force Q*** (the resistance due to the honey). [1 mark]

(b) (i) The speed is ***constant***. [1 mark]

(ii) This is because ***Force Q increases*** as the mintball accelerates, and eventually ***equals Force P***, which is constant. This means the ***resultant force is zero*** and so there is no acceleration. [3 marks]

(c) Between points B and C, the mintball travels 8 cm in 4 seconds. Use the formula ***speed = distance travelled ÷ time taken***, to calculate the speed is ***8 ÷ 4 = 2 cm/s***.
[2 marks]

Physics Paper 2 - Higher Tier

1 (a) (i) 13A - needs to be just a bit higher than normal working current. [1 mark]

(ii) The extra heat generated will melt the wire. [1 mark for wire melting/breaking, 1 mark for explanation]

(b) (i) $2.4 \times 3 =$ ***7.2 kWh*** [1 mark for 2.4×3, 1 mark for answer]

(ii) $7.2 \times 9 =$ ***64.8p*** [1 mark]

---

2 (a) He is wrong [1 mark]. [An extra mark for EITHER obtaining braking distance for two speeds, where one is twice the other OR saying that because the graph curves upwards (or increases exponentially), the braking distance increases by a greater proportion than the speed... or something along those lines.]

(b) 23 m [1 mark for line drawn on graph from 16 m/s, 1 mark for correct reading of scale between 22 m and 24 m]

(c) It's because the van travels an additional distance in the time taken to think before applying the brakes (thinking distance) [1 mark]

© 2002 CGP

(d) Braking distance increases significantly because tyres have less grip (less friction) on road. [1 mark for correct answer, 1 mark for reason]

Stopping distance — those examiners just love it...

---

3 (a) (i) acceleration = change in velocity ÷ time = 7 ÷ 50 = 0.14m/s^2
[1 mark for calculation, 1 mark for correct answer]

(ii) Kinetic energy increases, [1 mark] potential energy decreases. [1 mark]

(b) Work done = Force × Distance = 150 × 20 = 3000 Nm or ***3000 J***
[1 mark for formula, 1 mark for 150 × 20, 1 mark for answer in either unit]

(c) efficiency = energy out ÷ energy in = 3000 ÷ 6000 = ***50%*** or 0.5 or 1/2
[1 mark for formula, 1 mark for putting values in, 1 mark for answer]

---

4 (a) (i) radio waves or microwaves [1 mark for one correct]

(ii) ultraviolet or x-rays or gamma rays [1 mark for one correct]

(b) (i) infra-red [1 mark]

(ii) gamma rays [1 mark]

(c) ultraviolet, X-rays or gamma rays [2 marks for any 2 correct]

Electromagnetic waves have loads of uses — and they always come up in exams.
Make sure you know at least one use of each wave, and what order to put them in.

---

5 (a) The magnetic field will disappear [1 mark], and the paper clips will fall off [1 mark].

(b) (i) As the current gets stronger, the electromagnetic field gets stronger [1 mark] so it attracts the iron bar [1 mark], which rotates on the pivot and breaks the circuit (which stops current flowing) and trips the switch/spring [1 mark].

(ii) It can be reset by putting the iron bar back where it was [1 mark for something along those lines].

Oh look. Another electromagnetism question.
Wonder if it might be worth learning some of this stuff then...

---

6 (a) The turning force (moment) for Mildred is greater, but both are the same distance from pivot, so Mildred must be heavier. [1 mark for noting distance same, 1 mark all correct] (This all comes from Moment = force × distance from pivot.)

(b) 1.5 × Jane's weight = 1 × Mildred's weight = 1 × 700.
So Jane's weight = 700 ÷ 1.5 = ***466.7 N*** (to 1 d.p.) [1 mark for substituting into the formula, 1 mark for rearranging, 1 mark for answer (accuracy between 0 and 2 dp)]

(c) (i) You should draw the line of best fit with equal numbers of crosses either side, and it must go through (0, 300), since that's the most reliable point.
[1 mark for decent line]

(ii) 310 – 300 mm = ***10 mm*** [1 mark for reading off the total length somewhere in the range 308-312 mm, 1 mark for subtracting 300 from that length. So you'd get the marks if you read 308 mm off the graph and got 8 mm for the extension.]

© 2002 CGP

(iii) 100 × Tension = 500 × 10 so Tension = 5000 ÷ 100 = ***50 N***
[1 mark for substituting into the formula, 1 mark for rearranging, 1 mark for ans]

Tension... headache... tension... headache...

---

7 **(a)**

Switch up/down	Points X on/off	Points Y on/off	Lamp A on/off	Lamp B on/off	Lamp C on/off	Lamp D on/off
up	OFF	OFF	ON	ON	ON	OFF
down	ON	ON	OFF	OFF	ON	ON

[1 mark for every correct column.]

(b) **(i)** 3A (same current flows through all points along series circuit) [1 mark]
(ii) 0 A (no current flowing through this bit) [1 mark]

(c) **(i)** 2 V (potential difference shared between all components in series circuit)
[1 mark for 6 V ÷ number of bulbs, 1 mark correct answer]
(ii) 0 V (no current flowing through this bit, so no potential difference) [1 mark]

Well, you didn't think you'd get away without a good old circuit question, did you...

---

8 **(a)** Alpha particle: Helium nucleus [1 mark]
Can be stopped by paper, skin or similar [1 mark]
Beta particle: Electron [1 mark]
Can be stopped by thin aluminium or similar [1 mark]

(b) **(i)** Alpha radiation is more damaging to body tissues. [1 mark]
Couldn't be detected because it would be stopped by skin. [1 mark]
(ii) 8 min would be too short, as it would decay before doctors had time to detect it. OR It would be impossible to get the isotope to the hospital before most of it had decayed. [1 mark for either point] 8 years would be too long as the harmful radioactive substance would be in the body for too long, and would be dangerous as it could cause cancer. [2 marks, must mention danger]

(c) Half-life is time taken for mass of carbon-14 to decay by one half.
So after 5600 years it's down to half of original amount, after 2 × 5600 years it's down to a quarter, and after 3 × 5600 years it's down to an eighth.
So it's 3 half-lives old, or 3 × 5600 = 16 800 years old.
[1 mark for meaning of half-life, 1 mark for 3 × 5600 or mentioning that it has 3 half-lives' worth of decay.]

.... yeah, whatever.

---

9 **(a)** **(i)** Electrons move through the metal wire carrying charge [1 mark].
(ii) The solution contains charged ions [1 mark] which are attracted to the charged electrodes [1 mark]. Some ions gain electrons at one of the electrodes, while others lose electrons at the other electrode, so there is an overall transfer of charge through the solution [1 mark for expressing the idea of gaining / losing electrons].

© 2002 CGP

(b) You need the formula Q = I × t (or charge = current × time)
So t = Q ÷ I = 800 ÷ 10 = ***80 s***.
[1 mark for formula/method, 1 mark for correct answer]

Pfhhhmmmfhh...... more formulas......

---

10 (a) (i) S-waves [1 mark]

(ii) P-waves [1 mark]

(b) (i) Refraction [1 mark]. The density of the crust and mantle increases with depth [1 mark], so the speed of wave changes / the wave is refracted [1 mark for either of these points].

(ii) There is a sudden change in density [1 mark], so refraction occurs [1 mark].

---

11 (a) The wave is repeatedly reflected off the inner surface of the fibre optic as the angle of incidence is always greater than the critical angle [1 mark], so you get total internal reflection [1 mark].

(b) Any two of the following [1 mark each]:

- Can carry more information
- Much less interference
- More secure (can't be "tapped")
- Signals don't need boosting so often.

Fibre optics — another exam classic.

---

12 (a) Long waves diffract easily over hills so long wave radio will reach the house.

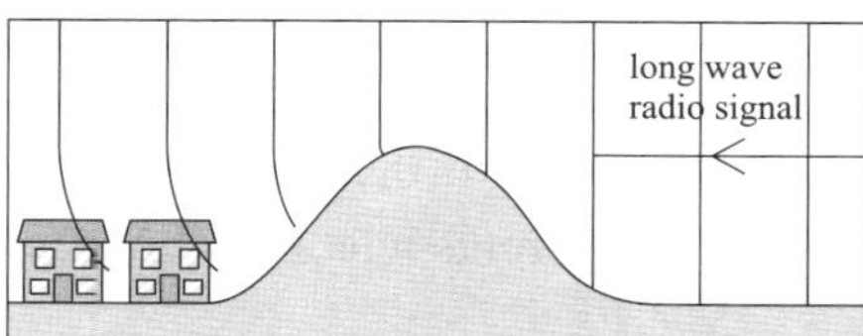

[1 mark for drawing waves reaching the house, 1 mark for bent ends, 1 mark for mentioning "diffraction"]

(b) Either: Short wavelengths aren't diffracted over hills, so FM won't reach the house.
Or: Short wavelengths aren't diffracted enough over hills for FM to reach the house.

No diffraction

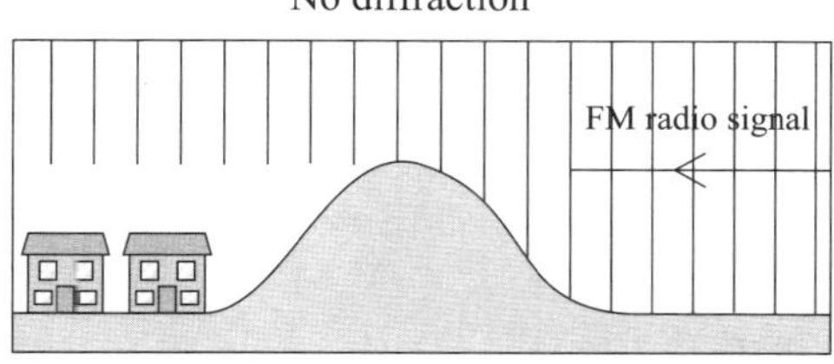

Little diffraction

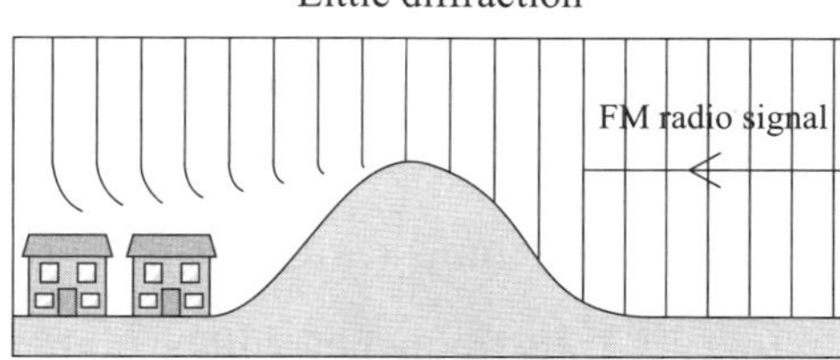

[1 mark for "shadow" area underneath waves, 1 mark for either of the above shapes of "shadow" area, 1 mark for saying there is no / little diffraction]

---

13 (a) Gravity [1 mark].

(b) Each planet orbits around the Sun [1 mark] in an elliptical orbit [1 mark].
The further the planet is from the Sun, the longer its orbit takes [1 mark].

© 2002 CGP

(c) Comets also move around the Sun in elliptical orbits, but their orbits are very eccentric (i.e. elongated), with the Sun close to one end [1 mark]. These eccentric orbits take the comets very far away from the Sun [1 mark], and it is only when the comets come much closer to the Sun that we see them [1 mark].

(d) Choose 2 from:
Communication, navigation, monitoring the weather, spying, space research (e.g. the Hubble Telescope), or any other reasonable answers [1 mark for each of 2].

Physics Paper 3 - Higher Tier

1 **(a)** **(i)** The water's heated by convection, so hot water rises to the top and cooler water sinks to be heated. Having the element at the top would mean cold water at the bottom not rising to the element. It also means you don't have to completely fill the kettle for it to work. [1 mark each for any of these points, up to a maximum of 2.]

(ii) Power = energy ÷ time (where the time is in seconds).
So power = 420 000 ÷ 3.5 mins = 420 000 ÷ 210 seconds = ***2000 W***.
[1 mark for power = energy ÷ time, 1 mark for remembering to convert to seconds (210), 1 mark for answer]

(b) Only the fastest water molecules evaporate, ie the ones with the most energy [1 mark]. That means the average energy of the remaining molecules will be lower [1 mark] — so the temperature of the liquid is reduced [1 mark].

(c) Air is heated by a heating element [1 mark]. This warm air then rises and colder air moves in to take its place [1 mark]. This 'new' air is then heated and begins to rise. All this movement of air sets up convection currents which eventually heat the whole room [1 mark].

(d) The fan increases the air circulation in the room. This means the flow of air over the element is increased and the room heats faster [1 mark].

Conduction, Convection, Radiation. Don't forget to learn a few real life applications so you can casually drop 'em in your answers and make it look like you really know your stuff.

---

2 **(a)** Radio waves, ***microwaves*** [1 mark], infrared, visible, ***ultraviolet*** [1 mark], x-rays, gamma rays.

(b) Gamma rays [1 mark].

(c) Its frequency will be higher [1 mark].

(d) Sound waves are ***longitudinal waves*** (not transverse waves like electromagnetic radiation) and so the ***vibrations / disturbances are in the same direction as the wave is travelling***. Sound waves also ***require a medium*** to travel in [2 marks for any 2 of the points in bold].

© 2002 CGP

3 (a) (i) First work out the number of half-lives:
1600 ⟶ 800 ⟶ 400 ⟶ 200, so 1 hour is 3 half-lives [1 mark].
Therefore 1 half-life must be ***20 minutes*** [1 mark].

(ii) One more hour is another three half-lives:
200 ⟶ 100 ⟶ 50 ⟶ 25.
So ***25 counts per second*** [1 mark].

(b) (i) Paper, skin (or other equivalent answer) [1 mark].
(ii) Thin aluminium (or other equivalent answer) [1 mark].
(iii) Thick lead, thick concrete (or other equivalent answer) [1 mark].

---

4 (a) (i) Acceleration due to gravitational force (or gravity) [1 mark].
(ii) He stops accelerating [1 mark], as the drag (air resistance and friction) increases with speed. Terminal speed achieved when gravitational force (in the direction of the slope) is exactly equalled by force of drag (air resistance and friction) [1 mark].
(iii) He needs to reduce his air resistance, so he needs to reduce his surface area or become more streamlined. He could crouch down low and/or wear aerodynamic clothing. He could reduce the friction of his skis. Extra weight would also help, as long as it didn't affect his aerodynamics/friction too badly. [1 mark for any one of these].

(b) (i) Distance = speed × time [1 mark], so distance = 22 × 7 = ***154 m*** [1 mark].
(ii) Decelerating (or slowing down) [1 mark].

Remember — the terminal velocity is when the force in the direction of the motion equals the drag (air resistance and friction). And the drag depends on the shape. Parachutists are the most likely example to come up.

---

5 (a) (i) There is no backward force on the runner, only the forward force A (from the runner's muscles). This means he accelerates [1 mark].
(ii) The bungee's exerting a backward force B, and this is larger than the runner's forward force C — so the overall force (and hence the acceleration) is backwards, and he slows down [1 mark].
(iii) The force D is much larger than the runner's forwards force E — so the backwards force (and the acceleration) is large [1 mark].

(b) Forward and backward forces were equal [1 mark].

Don't worry, you can forget all about this stuff in July.

---

6 (a) A: Force due to gravity (its weight) [1 mark].
B: The reaction force of the table [1 mark].

(b) (i) The teapot is not accelerating [1 mark].
(ii) They act in opposite directions [1 mark].

(c) A: becomes smaller [1 mark],
B: becomes smaller [1 mark].

Flying teapots? — not if Newton gets his way...

© 2002 CGP

7 (a) Points you could make:

Coal advantages:

(i) cheap, (ii) proven technology. (iii) easily controlled for changing demand

Coal disadvantages:

(i) pollution (acid rain, dust, CO_2, greenhouse gases),
(ii) coal mines messy (visual pollution), (iii) non-renewable resource.

Nuclear advantages:

(i) clean (eg no gas emissions), (ii) efficient, (iii) fuel cheap.

Nuclear disadvantages:

(i) dangerous process, (ii) dangerous waste, (iii) expensive,
(iv) not easily controllable for changing demand.

[6 marks for any 6 valid points].

Coal disadvantages: they make you shovel it with bare hands...

(b) (i) Wind; hydroelectric; wave; tidal; geothermal; solar; wood burning.
[1 mark for any one of these].

(ii) ***Wind advantages:*** no pollution; renewable.
Wind disadvantages: expensive; needs loads of ugly windmills; needs exposed site; no good when wind not blowing; low energy output.
Hydroelectric advantages: no pollution; renewable.
Hydroelectric disadvantages: needs mountains; environmental damage (flooding).
Wave advantages: no pollution; renewable.
Wave disadvantages: expensive; visually unattractive; needs waves; low output.
Tidal advantages: no pollution; renewable.
Tidal disadvantages: expensive; visually unattractive; restricts access to shipping; causes problems for wildlife; low energy output.
Geothermal advantages: no pollution; renewable; can use water directly for heating
Geothermal disadvantages: expensive; needs right kind of rocks.
Solar advantages: no pollution; renewable.
Solar disadvantages: expensive; only works well in sunny places; low output.
Wood burning advantages: cheap, proven technology, renewable
Wood burning disadvantages: trees take a long time to grow so forests need to be carefully managed, pollution (CO_2 etc); low output
[1 mark for a correct advantage and 1 mark for a disadvantage].

Solar panels — not much good in Cumbria...

---

8 (a) Energy = mass × SHC × temp change [1 mark].
Energy = 0.1 × 1100 × 30 [1 mark]. Energy = 3300 J [1 mark].

(b) Efficiency = useful energy transferred ÷ total energy transferred.
Efficiency = 3300 ÷ 18 000 [1 mark]. Efficiency = 18.3% (to 1 dp) [1 mark].

(c) Kinetic energy increases [1 mark].

(d) Thermistors are temperature-dependent resistors (i.e. their resistance changes with temperature) [1 mark] and so they will produce current changes in a circuit, which can be used to turn the elements on or off [1 mark].

© 2002 CGP

(e) So that the different electrical devices can be switched on and off separately, and if one thing stops working, the others continue to function [1 mark], and so that the full supply voltage is available for each item of equipment [1 mark].

---

9 ***Boiler:*** The boiler burns fuel, and so chemical energy in the fuel is converted to heat energy [1 mark]. This heat energy converts water to steam [1 mark]. ***Turbine:*** The steam drives the turbine, converting heat energy to kinetic energy [1 mark]. ***Generator:*** The rotation of the blades in the turbine drive the generator, making electricity. This converts kinetic energy to electrical energy [1 mark].

---

10 (a) A: Jupiter [1 mark],
B: Neptune [1 mark].

(b) They reflect light from the Sun [1 mark].

(c) (i) This means the Sun's gravity at Uranus is ***four*** [1 mark] times ***weaker*** [1 mark] than the Sun's gravity at Saturn.

(ii) Mercury [1 mark]. Because it's nearer the Sun, it has to travel more quickly to counteract the stronger gravitational force [1 mark].

---

11 (a) A step-down transformer [1 mark]. There are fewer turns on the output coil [1 mark].

(b) Nothing — transformers only work with a.c. [1 mark]. (Apart from an initial surge when the power is first switched on.)

(c) High voltage means low current [1 mark], and a low current means low power losses (because less heat is generated) [1 mark].

So, who's for A-Level Physics then?

---

12 (a)

Particle	Relative mass	Relative charge
proton	***1***	***+1***
neutron	***1***	***0***
electron	$1/2000$	−1

[1 mark for each particle completely correct].

(b) (i) C and D [1 mark].

(ii) Isotopes of an element have the same number of protons [1 mark] but different numbers of neutrons [1 mark].

(c) A beta particle is an electron [1 mark]. When a beta particle is emitted the charge in the nucleus of the atom increases by 1 [1 mark] and the mass number stays the same [1 mark]. Alternatively, give the mark for saying that a neutron changes into a proton and an electron (the beta particle).

© 2002 CGP

(d) **(i)** Beta particles [1 mark]. Alpha particles are stopped completely by paper; gamma rays would pass through any thickness of paper [1 mark].

(ii) A long half-life means that the radiation is basically unchanged over time, so the only change in radiation measured by the detector would be because of the thickness of the paper; a short half-life means that the strength of the radiation would change through time, even if the paper were the same thickness [1 mark for expressing this kind of idea]. If the half-life is too short, the radiation source would also have to be replaced more often [1 mark].

(iii) You'd need a different source because beta particles wouldn't pass through metal [1 mark]. A gamma source is needed [1 mark].

---

13 **(a)** Similarities: They travel at the same speed; they are both parts of the electromagnetic spectrum [1 mark for either of these].
Differences: Radio waves have a longer wavelength and a lower frequency than visible light. Also, radio waves cannot be seen by the human eye, whereas visible light can be. [1 mark for any of these].

(b) Long wavelength radio waves are diffracted around the hill [1 mark]; the shorter wavelength TV signals do not diffract as much [1 mark].

(c) **(i)**

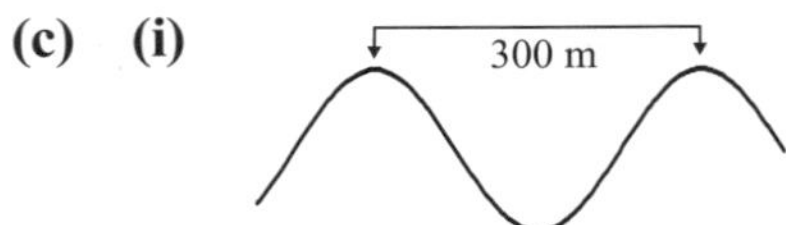

[1 mark for a diagram of wave showing crests and troughs, 1 mark for correctly labelling wavelength from crest to crest or trough to trough.]

(ii) Velocity = frequency × wavelength [1 mark],
i.e. 300 000 000 = frequency × 300 m [1 mark],
So frequency = 300 000 000 ÷ 300 = 1 000 000 Hz [1 mark].

(d) Sound waves are longitudinal waves, so the particles vibrate in the direction of travel of the wave [1 mark]. Radio waves are transverse waves, so the particles' vibrations are perpendicular (at right-angles) to the direction of travel of the wave [1 mark].
OR [1 mark for sound needs a medium i.e. air, whereas radio waves can travel through a vacuum]

Time for a cup of tea? Yeah, I reckon I deserve one.

PHP4U

© 2002 CGP